D1391131

The Age of BIG SCIENCE

1945 TO 1960

Published by The Reader's Digest Association Inc.
London • New York • Sydney • Montreal

transistors PHILIP
c'est plus sûr !

LIFE
A GENIUS AND HIS
MAGIC
CAMERA
Dr. Edwin Land of Polaroid
demonstrates his
new invention

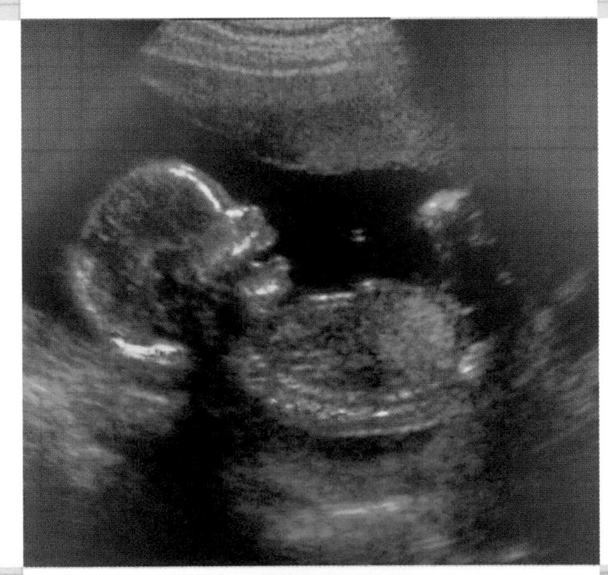
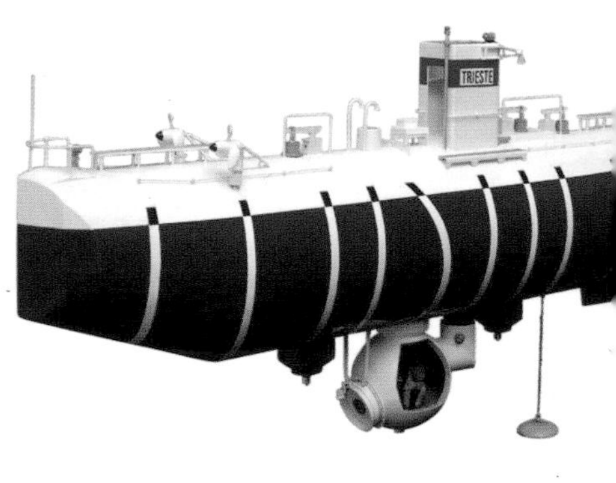

SONY
Betamax C7

Contents

The Atomium
The futuristic structure created for the Brussels World's Fair of 1958 became a symbol not just of the city but of the new forward-looking Europe.

Introduction

At the end of the Second World War, the overriding concern of world leaders was that carnage on this scale should never happen again. The United Nations was created with this agenda and the European Economic Community was born with the aim of reconciling Germany and France. Yet despite the rhetoric of peace, the two major powers, the USA and USSR, set about dividing the world into spheres of influence. As Winston Churchill put it, an 'iron curtain' descended across Europe.

The 1950s were about restoring global stability, although regional conflicts abounded. The Suez Crisis of 1956 confirmed the decline of Britain and France as leading players on the world stage. The Cold War unfolded against a backcloth of increasing repression in Eastern Europe and McCarthyite witch-hunts of suspected communists in the USA. Developing countries in Africa and Asia gradually began to free themselves from the yoke of colonialism and present a new non-aligned perspective on world affairs.

Capitalism and state socialism pursued very different economic paths, but for a time were evenly balanced: the USSR put the first satellite and first man in space, but the USA put the first men on the Moon. There were large-scale projects in housing, transport and other infrastructure, such as electricity generation – several countries began developing nuclear power to this end. There were also momentous changes in the fabric of society. Many more women gained the right to vote, and the Pill heralded a sexual revolution. The provision of welfare established safety nets against unemployment and ill-health, life expectancy increased and education became much more widely available.

Hand-in-hand with the economic progress of the US dollar came increased influence of American values and lifestyle. This was the age when America gave the world McDonald's, rock'n'roll and colour television.

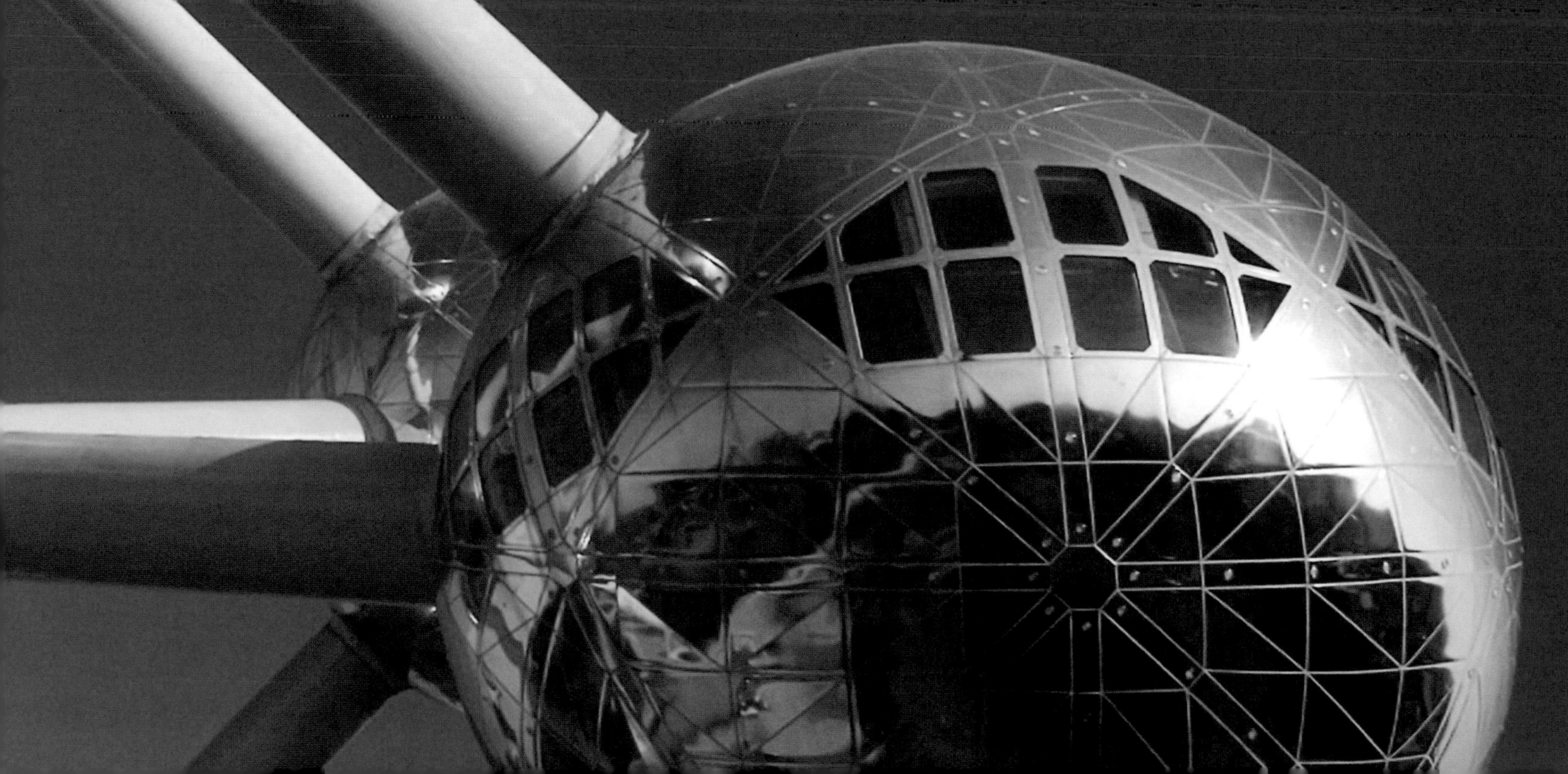

▼ The reliable, comfortable, economical, stylish and popular Vespa scooter made its debut in 1946; this model is from 2005

▲ Circuit boards from four generations of early computer

► The technique of radiocarbon dating using the carbon-14 isotope, devised by Willard F Libby, allowed archaeologists to pinpoint the age of remains that were once living matter

The end of the Second World War in 1945 saw the start of a period of unprecedented economic growth hand-in-hand with scientific and technological progress – a confident boom era that would last for the next quarter of a century. Two key inventions

► The Americans Bardeen and Brattain exploited the properties of semiconductors to create the first transistors in 1947; the increasingly small size and low energy demands gave rise to radios that could be taken anywhere

▲ Following in the footsteps of Ernest Rutherford, discoverer of the proton, scientists developed ever more powerful particle accelerators – from the cyclotron of 1931 (above) to the synchrotron, which in 1947 was the most advanced piece of equipment yet built for investigating the composition of matter

► The pioneering anthropological researcher, Claude Lévi-Strauss; in his work he attempted to dispel the concept of supposedly 'primitive' and 'civilised' cultures

ushered in the age of modern electronics: the computer and the transistor, which before long was being incorporated into integrated circuits. With the rise of the consumer society, labour-saving inventions came to play an ever greater role in daily life.

◄ In 1948, the Polaroid sparked a craze for instant photography that would be overtaken in the digital era; the company ceased production of its cameras and film in 2008

► The Big Bang theory, expounded by George Gamow and others, brought together earlier notions such as the 'primeval atom' and the constantly expanding universe; and would eventually be confirmed by the detection of cosmic microwave background radiation in 1964–5

▲ Holography, invented in 1947, allows images to be reconstructed in three dimensions; this hologram was created by George Lucas for the 2005 film *Star Wars III*

Kitchen roll, Tupperware, espresso machines, glass-ceramic cookware, pressure cookers and non-stick pans appeared in the kitchen. At the same time, desirable new toys like Barbie® dolls and Lego bricks stimulated an ever-expanding market for products

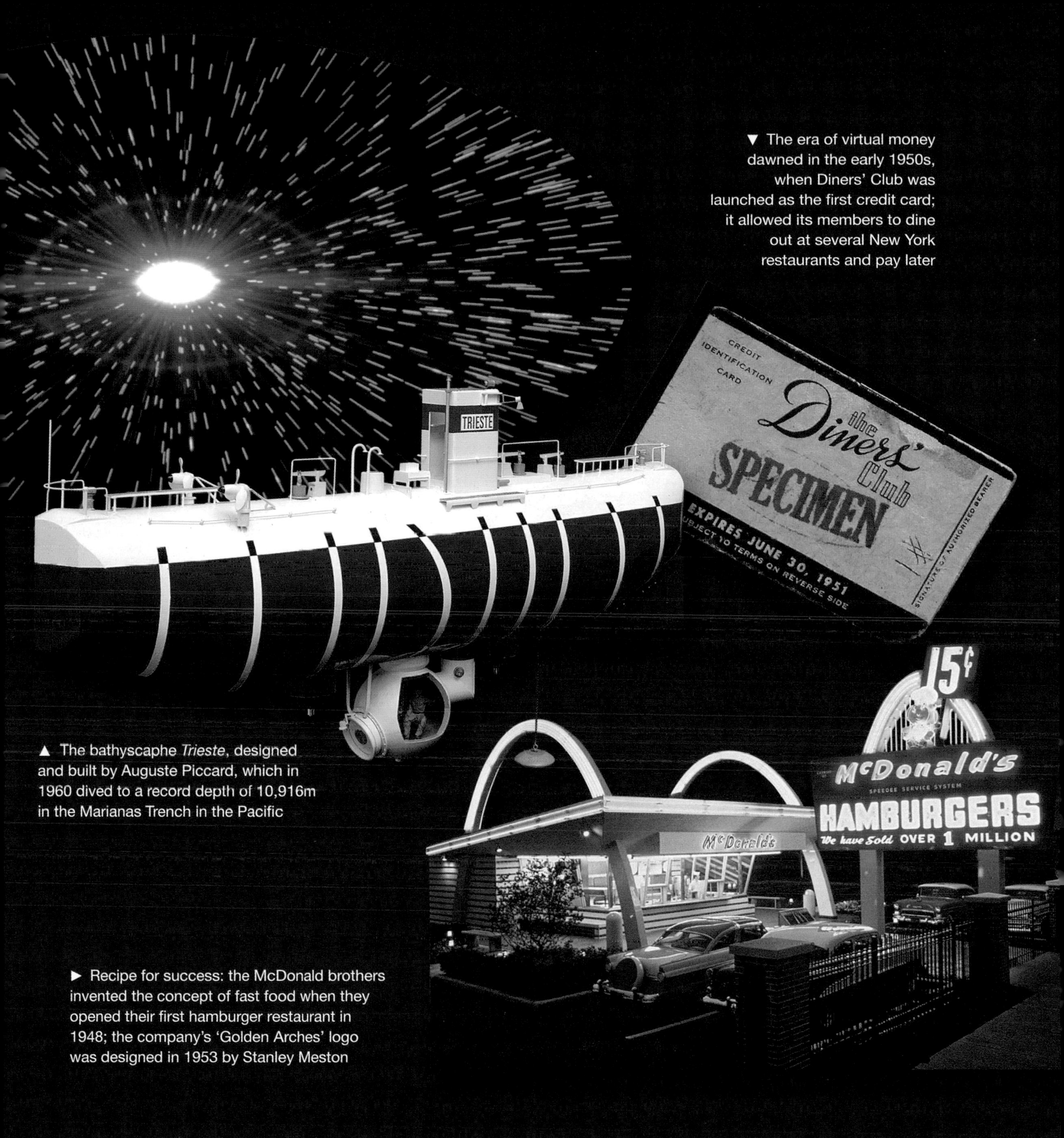

▼ The era of virtual money dawned in the early 1950s, when Diners' Club was launched as the first credit card; it allowed its members to dine out at several New York restaurants and pay later

▲ The bathyscaphe *Trieste*, designed and built by Auguste Piccard, which in 1960 dived to a record depth of 10,916m in the Marianas Trench in the Pacific

► Recipe for success: the McDonald brothers invented the concept of fast food when they opened their first hamburger restaurant in 1948; the company's 'Golden Arches' logo was designed in 1953 by Stanley Meston

aimed at children. Society was in a state of flux: the advent of the contraceptive pill, perfected in the 1950s but only readily available from the mid-1960s, changed the dynamic of sexual relationships, giving women more control over when – or whether – to start a

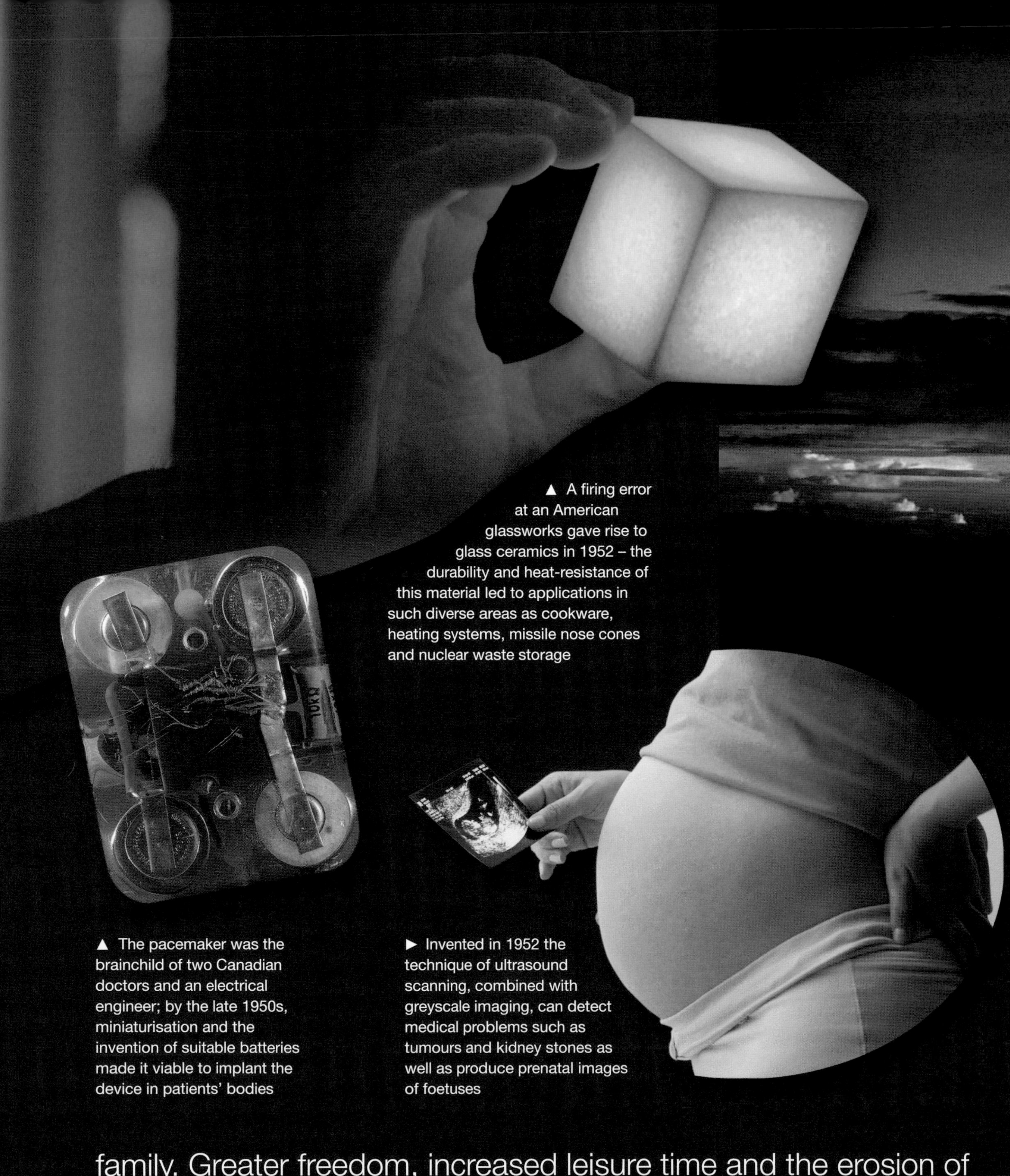

▲ A firing error at an American glassworks gave rise to glass ceramics in 1952 – the durability and heat-resistance of this material led to applications in such diverse areas as cookware, heating systems, missile nose cones and nuclear waste storage

▲ The pacemaker was the brainchild of two Canadian doctors and an electrical engineer; by the late 1950s, miniaturisation and the invention of suitable batteries made it viable to implant the device in patients' bodies

► Invented in 1952 the technique of ultrasound scanning, combined with greyscale imaging, can detect medical problems such as tumours and kidney stones as well as produce prenatal images of foetuses

family. Greater freedom, increased leisure time and the erosion of class barriers became the hallmarks of a popular aspirational culture that was catered to by such diverse phenomena as fast food, the Frisbee, the Polaroid camera and go-karting. The most

◀ The world's first hydrogen bomb was tested by the United States on 1 November, 1952, creating an explosion so powerful it created a 1km-wide hole in the ground below the detonation point

▲ The pressure cooker introduced a new fast method of cooking in 1954 and became so popular that entire cookbooks were devoted to recipes specifically for it

▲ In 1953 the discovery of the double-helical structure of DNA by Francis Crick (on the right) and James Watson revealed some of the mysteries of genetic heredity and paved the way for mapping the human genome, the creation of genetically modified foods and gene therapy

widespread of the new consumer goods was the television, later supplemented by the video recorder. Major advances in the medical field – such as ultrasound scans, pacemakers, anti-polio vaccines, the first organ transplants and kidney machines – would

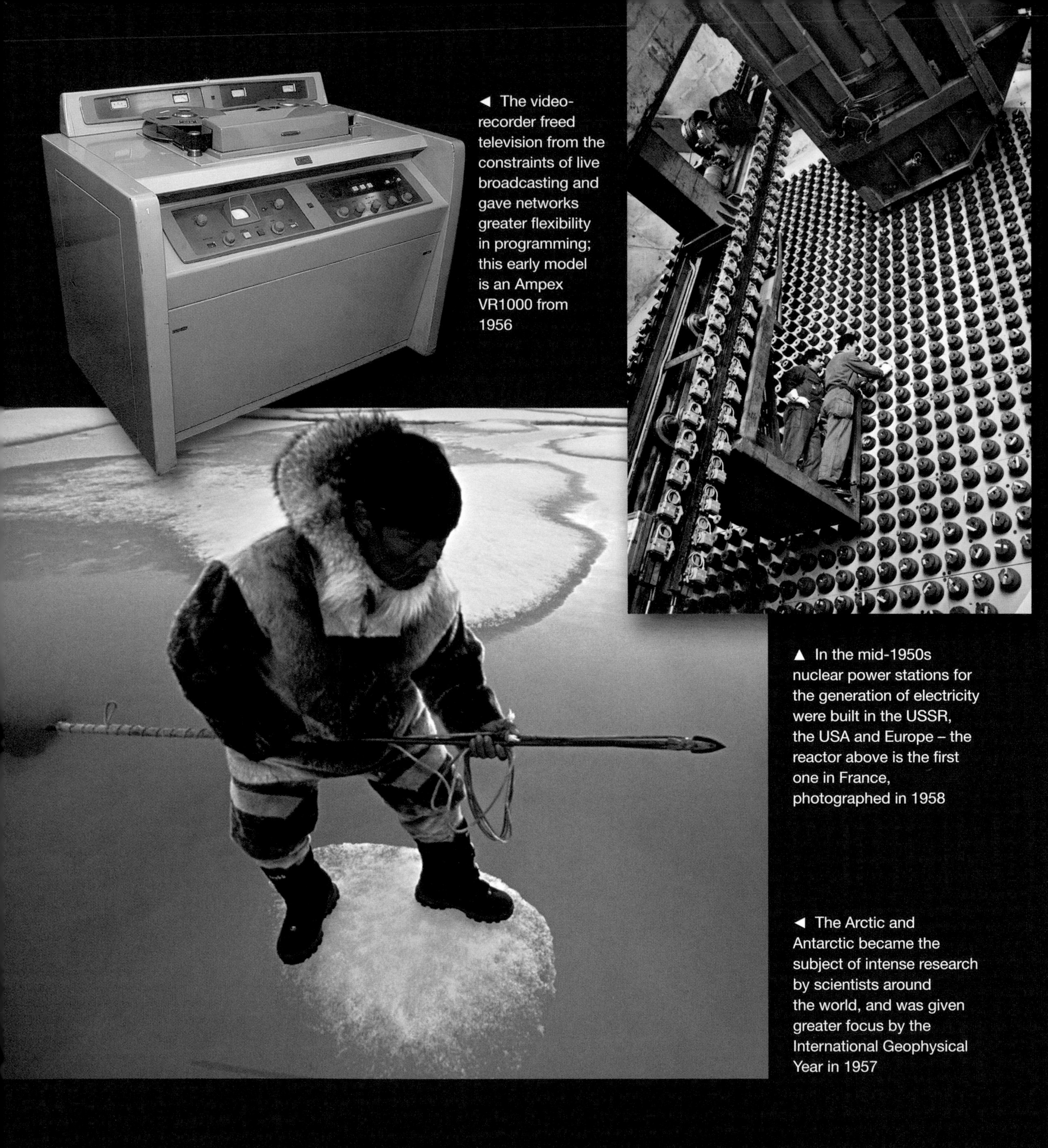

◄ The video-recorder freed television from the constraints of live broadcasting and gave networks greater flexibility in programming; this early model is an Ampex VR1000 from 1956

▲ In the mid-1950s nuclear power stations for the generation of electricity were built in the USSR, the USA and Europe – the reactor above is the first one in France, photographed in 1958

◄ The Arctic and Antarctic became the subject of intense research by scientists around the world, and was given greater focus by the International Geophysical Year in 1957

before long deliver steady improvements to people's quality of life. Medical research was given a further boost by the discovery of the structure of DNA and the ensuing development of molecular biology. From the microcosm of the human body to the

◀ In 1954 Jonas Salk's anti-polio vaccine was successfully tested in the USA and in 1988 world health agencies launched a programme to eradicate the disease around the globe

▼ 'Tefal' non-stick pans, introduced in 1956, were the invention of French aeronautical engineer Marc Grégoire

▲ The American surgeon Joseph Murray performed the first human organ transplant in 1954, when he transferred a kidney between identical twins; six years later he repeated the feat with non-identical twins

macrocosm of the Universe, fundamental physics – promoted by visionaries like George Gamow – also broke new ground with the Big Bang theory and the discovery of elementary particles. The onward march of nuclear power continued in the military realm

▼ Patented in 1955 by British engineer Christopher Cockerell, the hovercraft is a versatile amphibious craft that can traverse water, snow, sand and ice

▼ The world's first nuclear-powered submarine, USS *Nautilus*, was launched in 1954, the first submarine able to remain underwater for months at a time; by the 1960s, both superpowers had deployed a major part of their nuclear deterrent on missile-launching, nuclear-powered subs

▲ Fibre-optic cables transmit images in the form of light pulses and have found uses in medicine – endoscopes were invented in 1956 – as well as in modern telecommunications, transmitting digital data over long distances and at high speed

with the introduction of the hydrogen bomb and submarine-launched ICBMs. In the civilian world, the first nuclear reactors came on stream providing a new source of electricity. The world's oceans became a new frontier for exploration, investigated by the

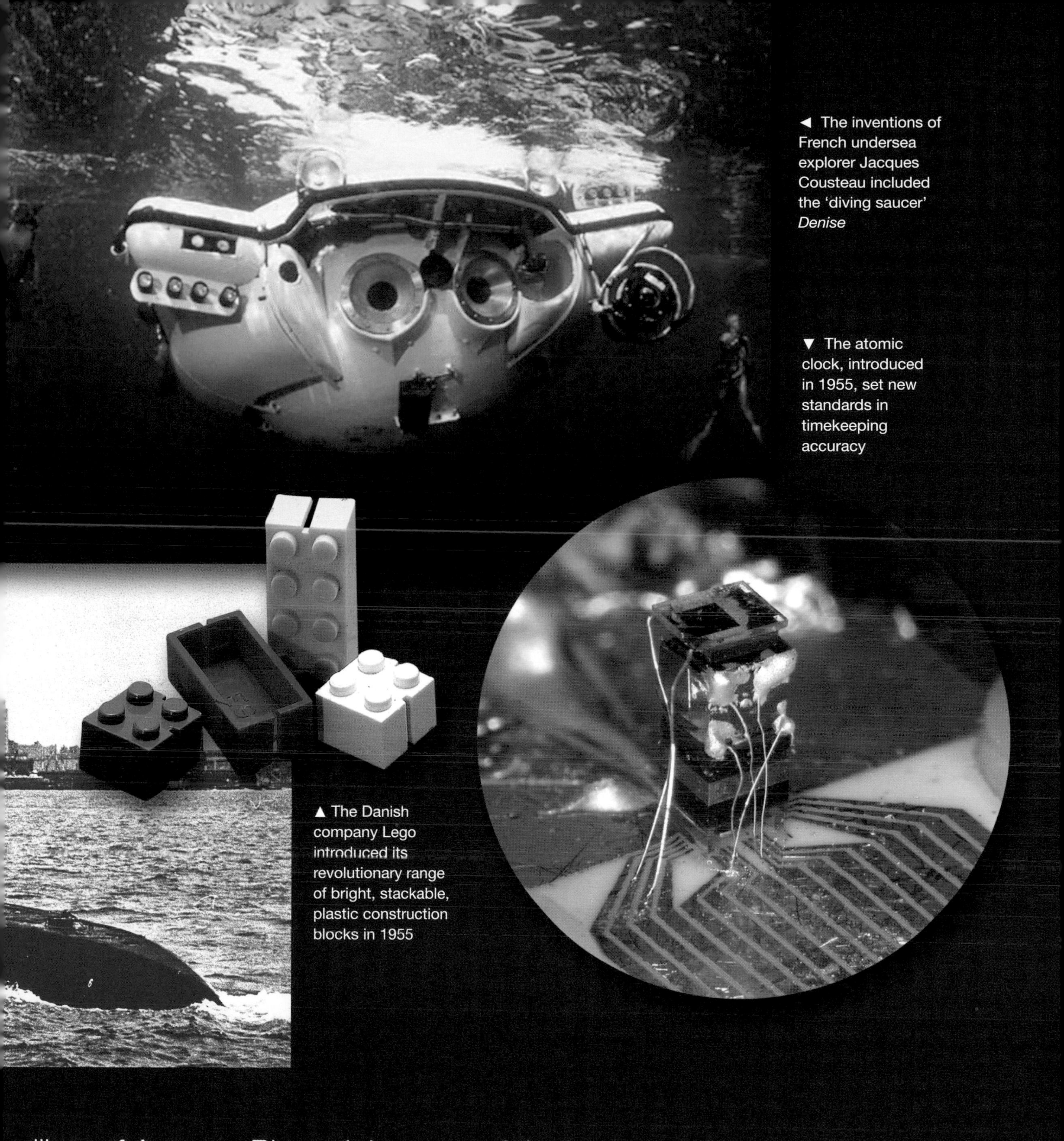

◄ The inventions of French undersea explorer Jacques Cousteau included the 'diving saucer' *Denise*

▼ The atomic clock, introduced in 1955, set new standards in timekeeping accuracy

▲ The Danish company Lego introduced its revolutionary range of bright, stackable, plastic construction blocks in 1955

likes of Auguste Piccard, inventor of the bathyscaphe, and the French adventurer Jacques Cousteau. Meanwhile, knowledge of the world's peoples was enriched by Claude Lévi-Strauss and his radical approach to anthropology. Economic output was in full

▲ The 1960s saw the introduction of the Pill, a new and highly reliable contraception that worked by inhibiting ovulation

► Television came of age with landmark broadcasts such as the coronation of Queen Elizabeth II in 1953 (right), the first televised US presidential debate between Nixon and Kennedy in 1960 and the Apollo moon landing in 1969

▲ In 1963 a team led by Ralph Plaisted became the first to reach the North Pole by land using motorised transport – Skidoos built by Bombardier of Canada

swing with far-reaching innovations including fibre optics, synthetic diamonds, computer-assisted design and manufacture, and the laser. By the late 1950s, after years of austerity, reconstruction and adjustment to a new world order, the West began finally to emerge

▼ The integrated circuit was invented by Jack St Clair Kilby in 1958; since 1965 the number of electronic components that can fit on a 3mm² silicon chip has grown from 30 to more than 6 million

▲ The Atomium was designed by Belgian architect André Waterkeyn for the Brussels World's Fair of 1958, the same year that the Belgian capital was selected as the seat of the European Economic Community

◄ In 1960 Theodore Maiman announced the development of the first laser, one of the most significant inventions of the postwar period which has found practical applications from surveying and eye surgery to weapons guidance systems

into the 'sunlit uplands' that Winston Churchill had promised would follow an Allied victory in the war. Epitomising this burgeoning confidence was Expo 58, the first post-war world's fair, fittingly held in Brussels, the city at the heart of a dynamic new Europe.

THE STORY OF INVENTIONS

While traditional industries in Europe strove to rebuild countries devastated by war, on the opposite side of the Atlantic, two important and closely allied new inventions arose that would change the world of work and leisure beyond all recognition: the computer and the transistor. Over just a few decades, they brought a shift in people's lives that was every bit as momentous as the agricultural and industrial revolutions: the era of information technology.

COMPUTERS – 1945

Birth of the 'electronic brain'

Without question, the most iconic and momentous invention developed in the latter half of the 20th century was the computer. Inspired by earlier adding machines and the fanciful notion of an 'artificial brain', researchers progressively refined both hardware and software, giving rise to our modern information society.

On 30 June, 1945, the University of Pennsylvania in Philadelphia published the rough draft of a scientific paper by the mathematician John von Neumann. Entitled 'First Draft of a Report on the EDVAC', it described the logical organisation and makeup of a high-speed automatic electronic calculating device, offering a comprehensive account of the architecture and operation of what would later come to be known as the computer.

Neumann's groundbreaking report owed much to the work of earlier American and British researchers, who over the preceding years had built machines that could perform incredibly fast calculations. The most famous of these early machines was ENIAC (the Electronic Numerical Integrator and Computer), which was commissioned by the US army in 1943 for the specific task of calculating artillery firing tables. Installed in a laboratory at the University of Pennsylvania, where it occupied several rooms, this 30-tonne behemoth contained almost 18,000 vacuum tubes and

Maths meets physics
John von Neumann (below, on the left) pictured with the physicist Robert Oppenheimer, director of the Manhattan Project, which was begun in 1942 to develop the atomic bomb.

A BRILLIANT MIND

John von Neumann was born in Budapest in 1903. From an early age, it was clear that he was an extraordinarily gifted mathematical prodigy. Endowed with a phenomenal memory, he shone during his university career both in his native Hungary and in Germany. In 1930 he was invited to study at Princeton University in New Jersey. Three years later, when the Nazis came to power in Germany, he settled permanently in the USA. As professor of mathematics at Princeton's Institute for Advanced Studies (IAS), he made important contributions to fields of study as diverse as mathematical logic, quantum physics, statistics and economic theory. During the war, he served as a consultant on several military programmes, including the Manhattan Project. He then turned his attention to high-performance computing and in particular investigating the shortcomings of ENIAC. Although he was the sole signatory of the 1945 EDVAC report, it is now thought that his work in this area was part of a wider collaborative effort.

March of miniaturisation
The circuit boards from four generations of early computers, demonstrating how electronic components were growing smaller. From left to right: ENIAC (built in 1943), EDVAC (1949), ORDVAC (1951) and BRLESC (1961).

THE FIRST COMPUTER GIANT

International Business Machines (IBM) was founded in Binghamton, New York, in 1896 as the Tabulating Machine Company. Up to the 1940s it was the main supplier of punched-card data-processing equipment to US firms and government agencies. Its machines were used on the Manhattan Project at Los Alamos. In the 1950s IBM took an early lead in developing digital electronic computers, building a network of mainframe computers (the SAGE Project) for the US Air Force for air defence.

could perform around 100,000 calculations a second. ENIAC and its British counterparts – the series of so-called 'Colossus' calculators built during the war at Bletchley Park to crack the secret codes generated by German cipher machines – are generally regarded as the world's first electronic computers. Their historical antecedents were mechanical devices built to calculate logarithmic tables (first devised by John Napier in 1614).

Yet these calculating machines, however fast, were only capable of performing simple arithmetical functions. Solving more complex problems – which involves the application of logic to work out the sequence in which tasks need to be completed – was beyond their scope.

The first true computer

Reprogramming early machines like ENIAC required a complete rewiring (Neumann had personally done this many times), so obtaining the answer to a given problem invariably took several days. It was this that prompted Neumann to write his epoch-making report. In it, he outlined how a machine could be equipped with a central processing unit – the CPU, now often simply called the 'processor' – which reads and executes the instructions of a program. EDVAC (the Electronic Discrete Variable Computer) was also the first computer on which data, including programs, could be input on magnetic tape. It performed a predetermined set of operations (a program)

DEBUGGING THE SYSTEM

A computer malfunction is commonly referred to as a 'bug', but few people realise that the term can be traced back to a real insect. On 9 September, 1947, the mathematician Grace Murray Hopper was working on the giant Mark II computer at Harvard University when the machine began experiencing problems. An investigation found a moth trapped inside, which had caused a short-circuit. On removing the moth, the operators announced that they had 'debugged the system' and the phrase stuck.

FATHER OF CYBERNETICS

The mathematician Norbert Wiener, born in Missouri in 1894, had completed his maths doctorate by the age of 18. After studying at Cambridge and Göttingen universities, he was appointed professor at the Massachusetts Institute of Technology (MIT), where he began investigating set and probability theory, with a particular interest in stochastic, or random, processes. In the Second World War Wiener refused to join the Manhattan Project, but he did work on developing automatic aiming and firing of anti-aircraft guns, which led him on to study communication theory. His 1948 work *Cybernetics: Or Control and Communication in the Animal and the Machine* broke new ground in the understanding of feedback mechanisms. His theory of control systems and communication soon found its way into the most diverse fields including biology, psychology, automation, computer science and even sociology. Wiener was always concerned by the ethical implications of his work and refused to accept any government funding for his post-war research.

Master of programming
Norbert Wiener teaching at MIT in 1958. Wiener advanced the notion that every action could be expressed in terms of a program.

that were stored within an internal memory, which also held the data to be processed. The machine's components were coordinated by another integral feature, a kind of clock called the control unit. Internal memory and control, plus versatility in enabling a variety of new programs for solving different problems to be simply loaded from tape, were what set EDVAC and other first-generation computers apart from earlier calculating machines.

It took four years before EDVAC was built. In the interim, Neumann's ideas about storing both programs and data in random-access memory (RAM), in those days consisting of modified cathode-ray tubes, were first put into practice by Manchester University's Small Scale Experimental Machine (SSEM, or 'Baby'), which began operating on 21 June, 1948.

The growth of artificial intelligence

It was no coincidence that the first computers appeared in the immediate aftermath of the Second World War. The skills of mathematicians had never been in such demand as they were during this conflict, with top-secret duties ranging from cryptanalysis to construction of the first atomic bomb. The vast military budgets put at their disposal allowed them to pursue the longstanding ambition of creating an artificial brain – a machine capable of reproducing human thought processes.

The idea was born from the convergence of two ancient strands of thought. First, the

Beyond control
Fear of intelligent machines was raised by Arthur C Clarke in the character of HAL, the malevolent spaceship computer brilliantly brought to life by Stanley Kubrick in the 1968 film 2001: A Space Odyssey *(below).*

THE LIMITS OF AI

For all its increasing versatility and sophistication, the computer will never be a match for the human brain. It is a simple processor of data that is unable to assimilate experiences, synthesise knowledge or express emotions, and therefore lacks the human capacity for imagination. Computers function in a strictly linear manner, whereas human thought creates analogies. In other words, computers can do only what they have been programmed to do.

age-old dream of creating life out of inanimate matter, an idea with a long tradition stretching from the myth of the golem in Jewish folklore to Mary Shelley's novel *Frankenstein* published in 1818 to the concept of robots in the early 20th century. The best that the pre-computer age could achieve in trying to realise this dream were mechanical dolls and similar automata. Second, a long line of thinkers – including the 13th-century Majorcan mystic Ramon Llull, the German philosopher Gottfried Leibniz (1646–1716) and the British logician George Boole (1815–64) – had inquired into the mechanisms of the human mind to try to deduce general laws that might form the basis of artificial intelligence. Computation theory, as this discipline came to be known, came of age in the 1930s and 40s, a period dominated by four key figures without whom computer science would probably not exist.

Four key computer visionaries

In 1936 the British mathematician Alan Turing came up with the idea of a universal machine that was theoretically capable of solving every logical or arithmetical problem that could be expressed in the form of algorithms – that is, a finite sequence of instructions leading to a definite answer. Thus was born the concept of the program. Around the same time, in the USA, Norbert Wiener was laying the foundations of what he would later call 'cybernetics' – the study of control and feedback mechanisms both in machines and in living organisms. In 1938 the American mathematician Claude Shannon showed that Boolean algebra and binary arithmetic could be used to simplify the arrangement of electromechanical relays used in telephone routeing switches. Conversely, he claimed that it would be possible to use arrangements of relays to solve problems in Boolean logic. This formed the starting point for information theory. The final piece of the jigsaw was Neumann's work on the architecture and working principles of future computers.

SHANNON AND BOOLEAN ALGEBRA

Born in 1916 in Michigan, Claude Shannon first came across Boolean algebra while studying at MIT. He was still a student when, in 1938, he published an article on the similarity between electromechanical relays and Boolean logic. A relay is basically an electromagnet operating one or two switches, each of which can be either open or closed (corresponding to 0 or 1 in the binary system). By combining a series of switches, it is possible to create a logic gate with several entrances and exits, which can perform three functions: OR (alternative), AND (addition) and NOT (negation).

Shannon was seconded to help with cryptanalysis of coded messages during the Second World War, so collating his ideas on computing language was put on hold until after the war. In 1948, in a book entitled *The Mathematical Theory of Communication*, he introduced the concept of quantities of information, in which the basic unit (0 or 1 in Boolean logic) became the binary digit or 'bit'. Shannon insisted that his ideas applied solely to exchanges between machines and were inapplicable to interpersonal communication, but his works were used to form the basis of both information theory and communication theory.

Computer linguist
Claude Shannon, the inventor of binary language for computers, in 1948

Transistorised technology
A circuit board for the ATLAS computer, developed jointly by Manchester University and Ferranti in 1962.

Keeping tally
UNIVAC I was built by the Remington Rand company and installed at the US Census Bureau in 1952. The set-up occupied more than 35 square metres of floor space.

More power *IBM's 'Stretch' computer, which first came on stream in the late 1950s, was so much faster than its predecessors it gave rise to a new unit of information – the octet, equivalent to 8 bits.*

CHESS MATCH OF THE CENTURY

On 11 May, 1997, in New York, the reigning Russian world chess champion Garry Kasparov was beaten by a computer, IBM's 'Deep Blue'. The previous year, in Philadelphia, Kasparov had triumphed in his match against the machine. This time it was the computer that won. Deep Blue was designed and programmed with the help of chess grandmasters. It comprised 256 separate processors working in parallel and in its upgraded 1997 version it could evaluate 200 million positions per second. The six-match game ended with two wins to one in favour of the computer and three draws. Kasparov's defeat was generally regarded as a freak, and put down to tiredness and psychological pressure. He requested a rematch, but IBM refused.

Kasparov versus computer *The match between Kasparov and Deep Blue, the first computer to defeat a world chess champion.*

Increasing efficiency

There ensued a search for ever greater computing power, a process that is ongoing today. In 1951 the UNIVAC computer, a vacuum-tube computer operated by the United States Census Bureau, could perform around 1,905 operations per second. By 1959, the IBM 7030 – an early transistorised supercomputer known as 'Stretch', which was installed at the Los Alamos atomic research facility to help develop the H-bomb – had reached a processor speed of over 1 MIPS

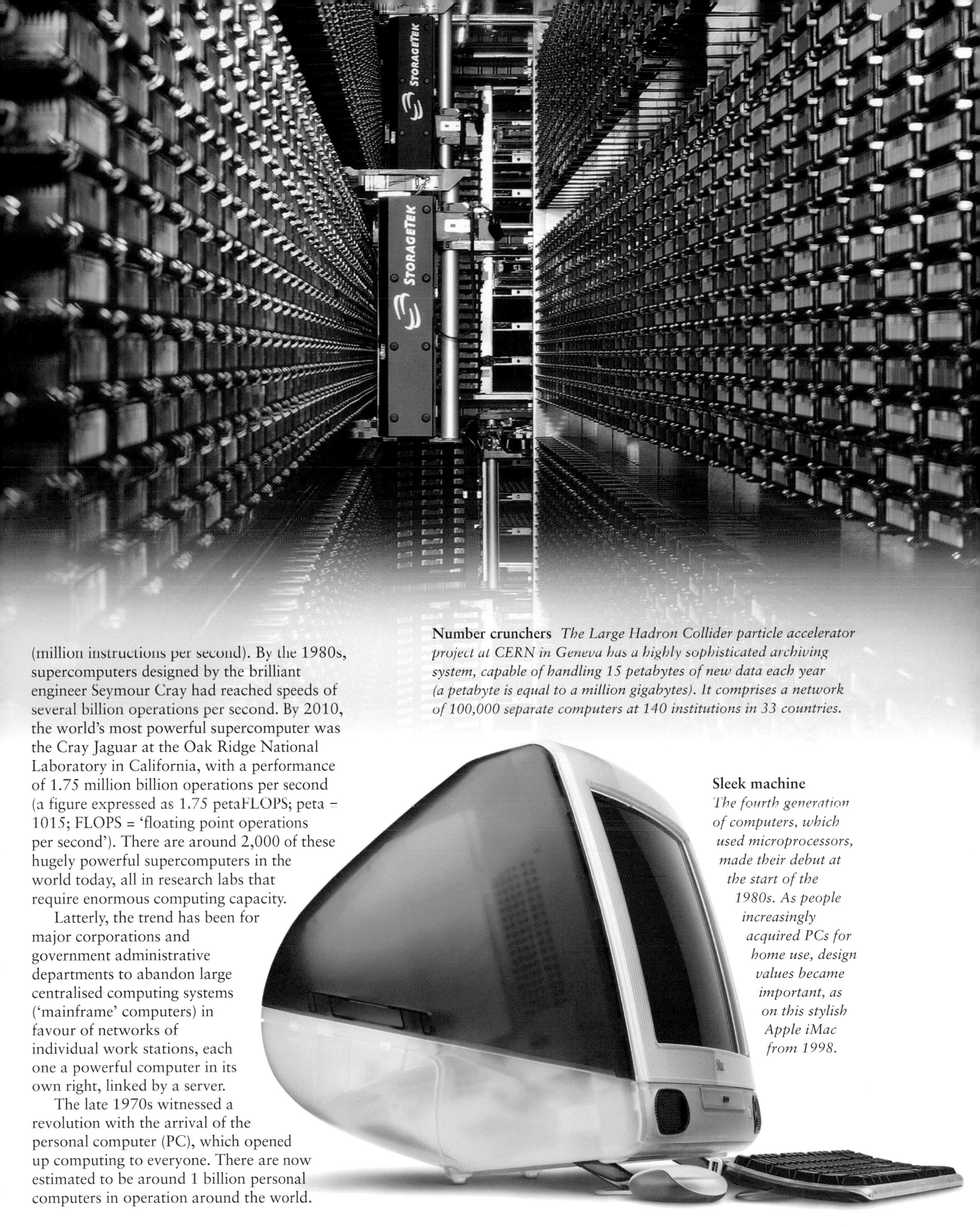

Number crunchers *The Large Hadron Collider particle accelerator project at CERN in Geneva has a highly sophisticated archiving system, capable of handling 15 petabytes of new data each year (a petabyte is equal to a million gigabytes). It comprises a network of 100,000 separate computers at 140 institutions in 33 countries.*

(million instructions per second). By the 1980s, supercomputers designed by the brilliant engineer Seymour Cray had reached speeds of several billion operations per second. By 2010, the world's most powerful supercomputer was the Cray Jaguar at the Oak Ridge National Laboratory in California, with a performance of 1.75 million billion operations per second (a figure expressed as 1.75 petaFLOPS; peta = 1015; FLOPS = 'floating point operations per second'). There are around 2,000 of these hugely powerful supercomputers in the world today, all in research labs that require enormous computing capacity.

Latterly, the trend has been for major corporations and government administrative departments to abandon large centralised computing systems ('mainframe' computers) in favour of networks of individual work stations, each one a powerful computer in its own right, linked by a server.

The late 1970s witnessed a revolution with the arrival of the personal computer (PC), which opened up computing to everyone. There are now estimated to be around 1 billion personal computers in operation around the world.

Sleek machine
The fourth generation of computers, which used microprocessors, made their debut at the start of the 1980s. As people increasingly acquired PCs for home use, design values became important, as on this stylish Apple iMac from 1998.

The computer

Computers were the successors to mechanical calculating devices. But there is more to a computer than the central processing unit: the other components and peripherals derive from a whole range of earlier inventions.

The smallest basic unit of digital information is the bit (binary digit), comprising the value 0 or 1.

A fragment of the 8 kilobyte fixed memory store from a Ferranti Atlas 1 computer of 1963.

LOGICAL OPERATIONS

FROM BOOLEAN ALGEBRA TO THE BIT

While ENIAC calculated in decimal base, the Manchester Mark I computer was the first to use binary notation – the bit – which is still the standard for modern computers. This numeral system, which employs just 0 and 1, goes back to the German mathematician Leibniz (1646-1716), who invented binary arithmetic. In the following century, in two seminal works, *Mathematical Analysis of Logic* (1847) and *An Investigation of the Laws of Thought* (1854), George Boole went further by showing that all logical operations can be described with the two values 0 and 1 and the three expressions AND, OR and NOT. In 1938 Claude Shannon showed that electric circuits with relays could be a model for Boolean logic, opening the way for Boolean algebra to become the basis of digital computer logic.

INTERNAL MEMORY

FROM THE CATHODE-RAY TUBE TO THE SEMICONDUCTOR CHIP

A computer's internal memory is used to store the data programme (also known as the 'instructions set') needed for current processing. The random access memory (RAM) in the Manchester Mark I (the first computer to store programs and data together in this way) used modified cathode-ray tubes (CRTs). The bits (basic units of information – 0 or 1) were stored in the form of charges on the inside surface of CRTs or as opaque spots on glass discs. The ferrite core memory, developed by Jay Forrester in 1953, was the norm until semiconductor memory appeared in the 1970s.

A Dynamic Random Access Memory (DRAM) unit, with over 1 million transistors (right)

ACTIVE COMPONENTS

FROM THE ELECTROMECHANICAL RELAY TO THE MICROPROCESSOR

Electromechanical relays, the first components that could assume two distinct states (in this case, open or closed), were replaced by vacuum tubes in the Manchester Mark I. But tubes were heavy, fragile and energy-inefficient as they heated up. The IT revolution only took off with the advent of transistors in 1947, which by 1954 were being made at low cost from silicon (a semiconducting material). The next major breakthrough came in 1958, when Jack St Clair Kilby and Robert Noyce of Texas Instruments created the integrated circuit. Instead of components being soldered onto a circuit board, they were now engraved directly into a thin substrate of silicon. In 1971 Intel unveiled the first microprocessor, the Intel 4004, which combined all the functions of the central processing unit (the CPU) on a single integrated circuit. Since then the basic technology has stayed the same, though the number of components per square centimetre on integrated circuits has grown exponentially.

The Intel 8080 microprocessor, launched in 1974 (right).

Integrated circuits installed on a printed circuit.

The JOHNNIAC of 1953, one of the first computers to incorporate von Neumann architecture (below).

Punched cards (right).

INPUT/OUTPUT

FROM PUNCHED CARDS TO PERIPHERALS

After the early days of IT, when computers had to be completely rewired in order to reprogram them, the first proper external input devices were magnetic tape units. Yet punched tape and punched cards for programming and other input, which dated from the era of mechanical calculators (the standard format for punched cards was set by Charles Babbage in 1888), still persisted until the arrival of the keyboard in the 1960s. A new kind of user interface was the mouse, unveiled in 1968. Output was initially in the form of magnetic tape, later replaced by perforated printout generated by fast teletype devices based on the electric typewriter. Computer screens, based on the oscilloscope, appeared in 1951. Data to be saved can now be downloaded onto various internal and external storage devices – hard disks, memory sticks, and DVDs.

The keyboard, a computer input peripheral.

STORAGE DEVICES

FROM MAGNETIC TAPE TO THE HARD DRIVE

The first computer storage devices were magnetic tape reader-recorders. The recording medium was a thin strip of nickel-plated bronze. Of the eight tracks on this tape, six were used for storing data. The first integral storage device was the drum memory installed on the ERA 1101 of 1950, which stored 1 million bits as magnetic pulses in tracks around a metal cylinder; read/write heads both recorded and recovered the data. The hard disk arrived in 1956 with IBM's Ramac 305. Its disk file consisted of 50 magnetically coated aluminium platters 60cm in diameter, stored one on top of the other, which could hold 5 million bytes of data. Modern hard drives are still constructed on the same principle, being made from aluminium or ceramics. Usually weighing less than a kilo, a hard drive can store huge amounts of data; in 2007 Hitachi introduced the first PC hard drive that could store more than a terabyte of data.

The magnetic tape unit for data storage on an Elliott 803, a small British computer of the early 1960s (left).

A computer hard drive from 1990.

COMPUTERS OF THE FUTURE

QUANTUM COMPUTERS AND INVISIBLE COMPUTING

As the limits of existing technology are reached, the next major development in IT will be 'quantum computers', which can harness the power of atoms and molecules to perform memory and processing tasks, and optical computers that use photons rather than electrons. Another innovation, already underway, is so-called 'invisible computing' – incorporating personal computers into everyday objects, effectively turning domestic appliances, cars and mobile phones into 'smart' technology. One idea, for example, is for streets to contain embedded computer chips that manage traffic and so enable cars to 'drive themselves'.

An Apple MacBook Pro laptop from 2006.

The Vespa scooter 1946

Italian style
Vespa scooters are still popular today – this model (right) is from 2005. They were pretty much an instant success: by 1950, just four years after its introduction, around 60,000 Vespa scooters were being sold annually. Vespa clubs sprang up around the world from 1952, a year in which the scooter got a publicity push from actress Audrey Hepburn, who rode one in the film Roman Holiday. *So celebrated a design icon did the Vespa become that the Spanish Surrealist artist Salvador Dali even customised one in 1962.*

After the collapse of Mussolini's dictatorship in 1945, Italy was banned from building aircraft. This had been the main activity of Piaggio, who now had to find something else to make. Brilliantly anticipating the huge surge in demand for personal transport in the post-war era, Enrico Piaggio asked his chief engineer Corradino d'Ascanio to come up with a modern redesign of the motorcycle, with the emphasis on making it fuel-efficient and, unlike pre-war motorbikes, as appealing to women as to men. In April 1946 the company lodged a patent for a series of revolutionary two-wheeled motorised vehicles and turned its factory over to this new production line.

Taking his cue from familiar aircraft construction techniques, d'Ascanio completely revitalised the concept of the two-wheeler. He attached each wheel to a single cantilevered oleo leg (as on a plane's undercarriage) rather than using forks, which made it far easier to repair punctures or change tyres. A streamlined engine was mounted directly on the vehicle's back axle, which it drove like a propeller, dispensing with the traditional motorbike drive chain, which was dirty and inefficient.

A wasp buzzes round the world

D'Ascanio's radical design enjoyed instant success. The Piaggio company called their new product the *Vespa* (Italian for 'wasp', from the high-pitched buzz of its engine). In English-speaking countries it became known as the motor-scooter, because its small wheels reminded people of children's scooters. As the two-stroke engine was rear-mounted,

SCOOTER HERITAGE

An electric-powered two-wheeler was made as early as 1900 by the French company of Clerc and Pingault. It was followed two years later by the *Auto-Fauteuil* (literally 'auto-armchair'). In 1915, the 'Autoped' appeared in America, a vehicle like a motorised child's scooter, which was ridden standing up. Six months before the first Vespa was produced in Italy, the Fuji organisation in Japan launched a similar-looking vehicle called the 'Rabbit'. Both this and the Vespa had been inspired by small military scooters made by the Cushman company of Nebraska for use by personnel on US air bases in the Second World War. Their machines could be dropped with airborne forces and they became a familiar sight on the roads of occupied Italy and Japan in 1945.

VESPA VERSUS LAMBRETTA

One year after the launch of the Vespa, a home-grown rival emerged in the form of the Lambretta, manufactured by Innocenti in Milan (the name came from the River Lambro, which ran close to the factory). Innocenti originally specialised in making steel tubing but soon turned most of its production capacity over to scooters. As small, more affordable cars came on the market, scooter sales declined and the Innocenti factory closed in 1972. An Indian company bought up the plant and transferred production to Lucknow.

this left a platform free for the feet, so women could ride in skirts, while men could wear smart business suits without soiling them – a streamlined fairing at the front provided further protection for the rider. Plus, the scooter's manoeuvrability made it ideal for weaving through traffic.

The Vespa's affordability and modern styling soon made it a hit with Italian youth, and it became a symbol of a carefree urban lifestyle, appearing in many Italian films of the period. These helped to boost scooter sales worldwide and the 1950s saw Italian scooters being manufactured under license in Germany, Britain, France and Spain. By 1953, some 10,000 franchises in countries as far afield as the USA, Brazil, Iran and China were selling Vespas. From the 1960s to the end of the 20th century, they were the most widespread form of personal transport in India and South America. In many Asian countries, scooter sales still outstrip those of cars. In Britain scooters became firmly identified with the urban 'Mod' culture of the 1960s.

The durable scooter

Although Japanese motorbikes and mopeds began to make major inroads into the scooter market from the mid-1970s, thousands of customers remained loyal to Vespa scooters, valuing their rugged reliability. Scooters were relaunched in the 1980s with revamped styling and technological improvements. Automatic transmission, for instance, first fitted to Japanese scooters, became the norm. The noisy two-stroke engine, with its acrid, polluting exhaust fumes, was replaced by four-stroke models. Recent innovations include three-wheeled scooters (with two wheels up front for greater stability and easier parking) and fuel-efficient electric scooters.

Rugged workhorse
The scooter's carrying capacity is evident from this photograph of a young couple on a European camping holiday in the 1950s.

Family runabout
In many Asian countries, such as India, China, and Vietnam (right), scooters are utility vehicles rather than style icons, used for transporting goods or, as here, the whole family.

Artificial snow 1946

Snow man
Vincent Schaefer (far left, centre) creating artificial snow in a freezer. The US military later used the cloud-seeding techniques he developed to trigger rainfall in Vietnam with the aim of hampering the ability of the North Vietnamese to supply its army.

During the Second World War, the American chemist and meteorologist Vincent Schaefer succeeded in creating artificial snow in the laboratory by blowing air that was full of water vapour through a domestic freezer at a temperature below 0°C. He repeated the feat on 13 November, 1946, only this time in the open air, creating snowfall by 'seeding' the top of a cloud with dry ice (liquid carbon dioxide, which solidifies in contact with the air). Thus began a long series of experiments with cloud-seeding, with the aim of stimulating precipitation during periods of drought – or alternatively of causing violent storms to break before they reach cultivated areas and destroy the crops. The technique was not entirely successful: it was costly, involving as it did aircraft hire, unreliable and also turned out to be environmentally damaging as a result of the harmful chemicals used – in particular sodium iodide.

SNOW CANNONS

From the 1970s onwards, Schaefer's methods began to be employed on a grand scale to produce snow for pistes at ski resorts that had insufficient snowfall before the start of the season. The organisers of the 2010 Winter Olympics in Vancouver were forced to call on the technique. But skiers prefer natural snow, which offers much better traction.

Tupperware 1946

In 1946, an American chemical engineer devised a product that would quite literally make him a household name. Earl Tupper took polyethylene, a light plastic that was odour-resistant and unbreakable, and moulded it into bowls and other kitchen containers to create 'Tupperware'. A key feature of his range was the press-shut lid system, which made the bowls and tubs airtight and stackable. Yet the product only really took off in 1948, when a sales representative called Brownie Wise had the bright idea of the 'Tupperware party'. In return for a percentage of sales, housewives were invited to host get-togethers in their own homes to sell the containers to their friends and neighbours. By 1983 this direct marketing formula had taken Tupperware to six western European countries, as well as Australia and Japan. Today, the company has an independent sales force of more than 2 million worldwide and its products have moved with the times, being fully freezer-proof and microwaveable.

Tupperware takes off
Women inspecting the wares at a Tupperware party in 1951. This form of direct marketing was pioneered by the Tupperware company. The product styles are constantly evolving and examples have even been exhibited in galleries of modern and applied art, such as the Metropolitan Museum of Art in New York.

Kitchen roll 1946

In 1946 Stanslas Darblay, the descendant of a French papermaking family, went on a research trip to the USA. There he discovered a remarkable type of paper that looked like cloth and mopped up liquids just as efficiently. Back in Paris he put his discovery to work and created Sopalin, the world's first manufactured kitchen roll. The very first batch of kitchen towels had been the result of quick thinking by Arthur Scott back in 1931, when a batch of paper delivered to the Scott Company of Philadelphia was discovered to have been rolled out too thickly for toilet tissue. The kitchen roll popular today, often made from recycled paper, did not make an impact until the 1960s.

Easy cleaning *A French advertisement from the 1960s extols the virtues of Sopalin kitchen roll, which did away with the need for potentially unhygienic cleaning cloths around the kitchen.*

The espresso machine 1947

One of the stars of the Paris World's Fair of 1855 was a hydrostatic percolator, invented by Édouard Loysel de Santais. Crowds flocked to watch the machine as it produced large quantities of piping-hot coffee. Others followed, producing a bewildering variety of weird and wonderful coffee machines. They all operated on the basic principle of forcing boiling water and steam at high pressure through dark roasted, finely ground coffee.

Giovanni Achille Gaggia, a bar owner in Milan, came up with the definitive design for the espresso machine in the 1930s. Dissatisfied with the flavour of coffee from existing percolators, in which steam coming into contact with the coffee grounds made them bitter, he began experimenting with a rotary screw-type piston. In 1947, after much trial and error, he patented a spring-powered lever piston that eliminated the need for steam. The result – after just 15 seconds – was a strong coffee with the now-familiar frothy layer on top. Gaggia had found the ideal balance between temperature, pressure, quantity of coffee and percolation time to produce the perfect cup of espresso.

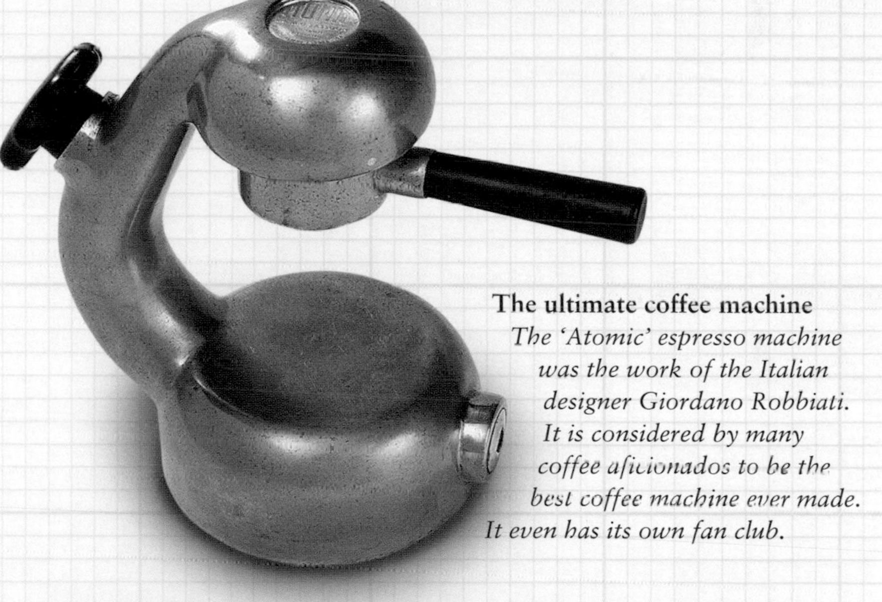

The ultimate coffee machine *The 'Atomic' espresso machine was the work of the Italian designer Giordano Robbiati. It is considered by many coffee aficionados to be the best coffee machine ever made. It even has its own fan club.*

GETTING IT JUST RIGHT

To make a cup of coffee that was not too bitter and with no burnt taste, Gaggia had to strike the right balance between speed and heat in the brewing process. Too much heat would impair the aromatic compounds that give coffee its flavour; too slow a percolation would make it bitter. Gaggia's lever-operated machine delivered just the right silky-smooth, rich taste.

RADIOCARBON DATING – 1947

Pinpointing the past

The introduction of a dating technique based on measuring the concentration of radioactive carbon isotopes in specially collected samples furnished archaeologists with an invaluable new tool. Radiocarbon dating is now used in a wide range of areas of study.

Precision work
Accelerator mass spectrometry (AMS), which began in the 1980s, helped to refine the carbon-14 dating method. The inventor of the radiocarbon dating technique, Willard F Libby (below), used a Geiger counter of his own devising in his experiments.

Before 1947, the only way for archaeologists to date their finds was the chronological method of stratigraphy. This technique was based on the premise that the lower strata of material at a site are older than those nearer the surface. Yet this approach, known as 'relative dating', could not determine the precise date of remains. Then, in 1947, Willard Frank Libby, a chemist at Chicago University, announced that he had found a method of absolute dating based on the radioactive isotope carbon-14, discovered in 1940. The technique involved measuring the concentration of this isotope in samples.

Carbon-14 (14C) occurs naturally in the Earth's atmosphere. It is absorbed by plants in the form of carbon dioxide (CO_2) and fixed during photosynthesis. Thereafter it enters every living creature through the food chain. In contrast to carbon-12 (12C), the stable isotope of carbon, 14C decays very slowly within the organism; it has a half-life of 5,730 years. While an organism is still living, its levels of 14C are constantly replenished from its food intake, but from the moment of death that external supply ceases and the concentration of the radioisotope within the organism begins to decrease. By determining the proportion of 14C compared to 12C still remaining in a sample, scientists can calculate how much time has elapsed since the organism died and thus give it a date.

Libby's method of gauging the 14C content involved burning some of the sample in order to convert it to carbon dioxide.

Lying in state
The coffin of Hor, a priest who lived during the 22nd dynasty in ancient Egypt (c945–715 BC), now in the British Museum. Radiocarbon dating is used to date burials if other pieces of immediate evidence, such as burial practices and inscriptions, prove inconclusive.

EXTENDING THE TECHNIQUE

Since Libby's time, radiocarbon dating has been further refined by the use of mass spectrometry of samples in particle accelerators. This can be used to date minuscule samples containing just a few milligrams of carbonised material, rather than the several grams required before. But there remains one major limitation: carbon-14 cannot date samples more than 50,000 years old, since their level of radioactivity is too low to measure. In these cases, scientists measure the concentration of other atoms, such as those of lead resulting from the decay of uranium, a method introduced by Bertram Boltwood in 1915. There are now around 20 different radiometric dating tests enabling scientists to date objects up to a billion years old. But unlike radiocarbon dating, which can date organic remains, these other methods can be used only with mineral samples.

Time lord
Willard Frank Libby (left) was honoured for his invention of radiocarbon dating with the award of the Nobel prize in chemistry in 1960.

A radiation detector (a hydrogen-filled Geiger counter) then counted the electrons given off by the decaying 14C.

Testing the theory

From 1949 onwards, Libby conducted practical trials of his dating method by analysing, among other artefacts, fragments of cypress and acacia wood from the sarcophagi

of two pharaohs from ancient Egypt, Snefru and Djoser. The answer he arrived at was 2650 BC (with a margin of error of plus or minus 75 years) and this tallied exactly with the regnal dates of these kings, which were already known from papyrus records.

Radiocarbon dating was eagerly adopted by archaeologists and palaeontologists. After Libby dated the prehistoric paintings in the Lascaux caves in France in 1951, placing them at around 15,500 BC (with a margin of error of 900 years), the technique became standard practice in both disciplines. It shed new light on countless finds from prehistory and classical antiquity, including not only human and animal skeletons and teeth but also any material that had once been living matter, such as textiles, paper, parchments, shells, charcoal, and feathers. Long-running historical controversies were also investigated, such as the age of the Turin Shroud (supposedly Christ's burial cloth), which was shown to date only from the Middle Ages.

A wealth of uses

Carbon-14 dating has proved an invaluable tool for investigating how the Earth's climate has changed over time. By measuring the 14C content in sediments from the bottom of oceans and lakes, and collating this data with the presence of particular pollen grains or micro-organisms, it is possible to reconstruct an exact chronology of the successive climatic events that have shaped the planet, and to determine key stages in the evolution of life forms. This also helps climatologists predict future patterns. Radiocarbon dating has enabled meteorologists to track the gradual decline in radioactivity in the atmosphere, following the overground nuclear testing of atomic and hydrogen bombs that took place from the late 1940s until 1963. And in agribusiness and the perfume industry the technique is used to test the quality of substances in products that claim to be 'natural', as they can only be regarded as such if they yield the same 14C count as found in the current atmosphere. A gang of fraudsters was caught in this way when they tried to pass off inferior wine from the year 2000 as fine vintage claret more than 100 years old.

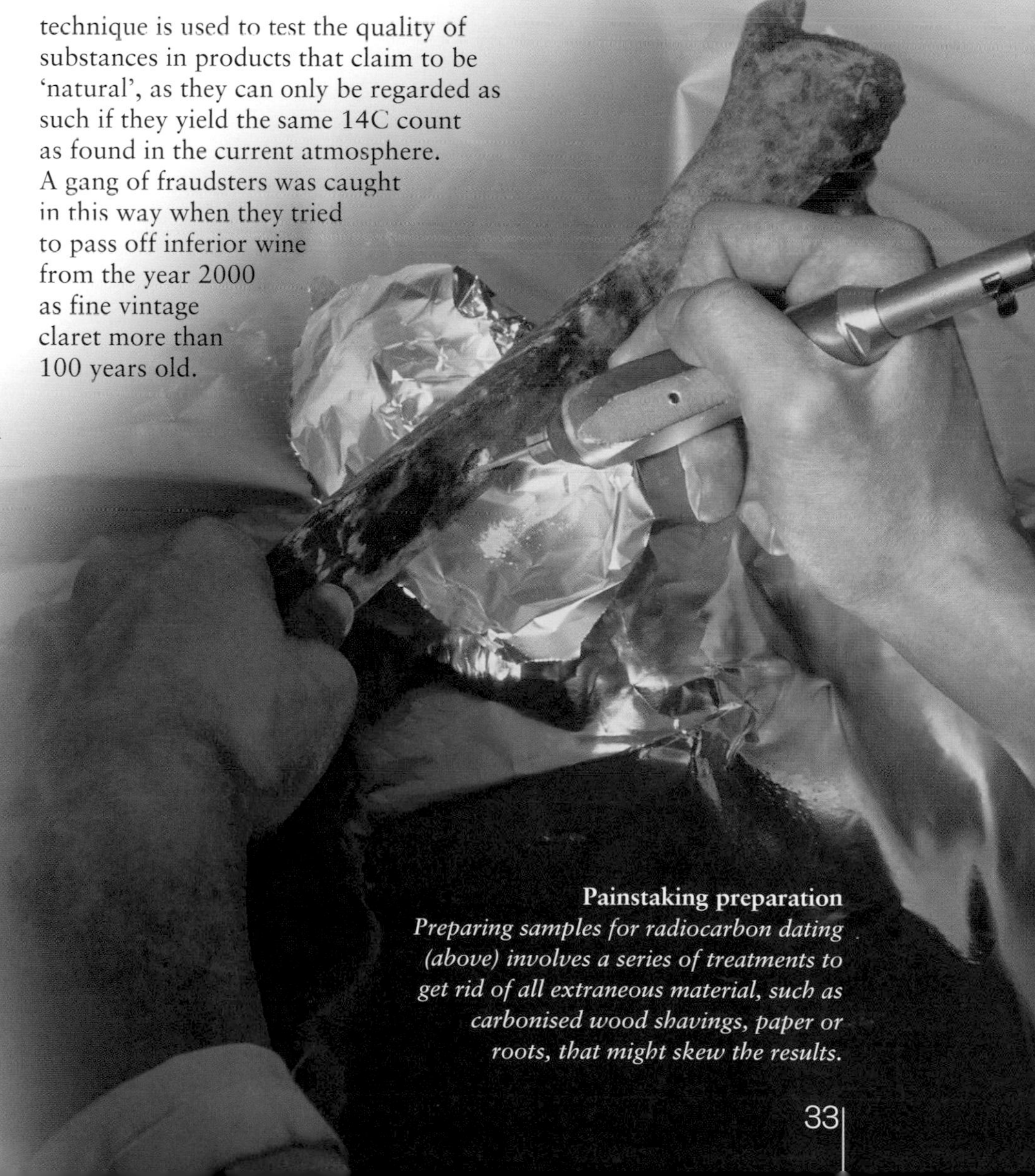

Painstaking preparation
Preparing samples for radiocarbon dating (above) involves a series of treatments to get rid of all extraneous material, such as carbonised wood shavings, paper or roots, that might skew the results.

Smashing the atom

The ancient Greeks believed that atoms were the smallest constituent elements of matter, a belief that lasted for two millennia. In the early 20th century physicists found that atoms were themselves composed of smaller entities – neutrons, protons, electrons – which in turn are made up of infinitesimally small particles known as quarks. One machine above all helped scientists to reveal the 'Russian-doll' structure of atoms: the particle accelerator.

Revolutionary apparatus
The first operational cyclotron (below), built by Ernest Lawrence at Berkeley in 1931, measured just 13cm in diameter.

In 1919, in Manchester, the New Zealand-born British physicist Ernest Rutherford achieved a major breakthrough with his discovery of the proton, one of the fundamental particles that make up the nucleus of the atom. Rutherford had predicted the existence of this particle several years before; now he set about gaining conclusive proof by bombarding nitrogen gas with helium nuclei (which he called 'alpha particles') emitted spontaneously by radioactive radium. Every now and then, he noted that the alpha particles would strike the nuclei of the nitrogen atoms and knock out a positively-charged particle – the proton (a term he coined the following year). This was the first atomic particle ever to be observed.

But almost as momentous as the discovery itself was the experimental method that Rutherford used to make it: particle acceleration was to become a vital element in the ever more important science of nuclear physics.

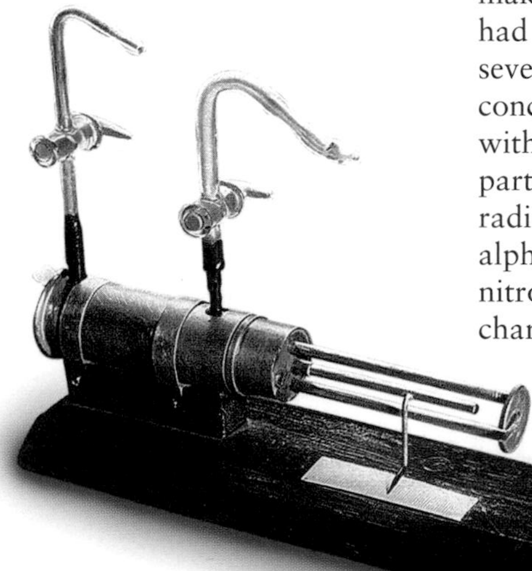

Crude yet effective
The apparatus (above) with which Ernest Rutherford first observed artificial transmutation in 1919: when nitrogen atoms collided with alpha particles from a source inside the horizontal enclosed tube, they were converted into oxygen atoms. When the nitrogen atoms transmuted into oxygen atoms they ejected protons, which were detected at the rectangular window at the end of the tube.

Speeding up the proton

The helium nuclei that Rutherford used were low-energy projectiles. In the wake of his historic experiment, the holy grail for physicists became particles that could be accelerated *ad infinitum* to heighten the force of the impact. The proton seemed the ideal candidate. Its positive charge meant that it was susceptible to an electric field, which could be used to accelerate it still more.

In Cambridge in 1932, John Cockcroft and Ernest Walton created the first linear particle accelerator, fitted with a voltage multiplier that built up a potential difference of 700,000 volts. They used it to bombard and disintegrate lithium nuclei with high-energy protons, but because their multiplier was at the furthest limits of generator technology, this kind of particle accelerator was a scientific cul-de-sac.

Making history
An artist's re-creation of Ernest Lawrence adjusting his experimental cyclotron (below, in the blue coat). Lawrence was awarded the Nobel prize for physics in 1939 for his work on the machine.

Around the same time, the American physicist Ernest Orlando Lawrence was exploring a different approach to high-energy particle physics. His idea was to try to prolong the period during which protons were under the influence of the electric field, a phenomenon known as 'phase stability'. He found that a weak current, if applied for a long time, could generate far greater particle velocity and energy than a momentary burst of high voltage current. The result was a circular accelerator that boosted the speed of the protons with every turn – the cyclotron.

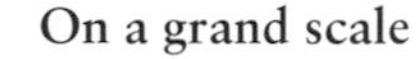

On a grand scale
The torus of the Large Hadron Collider at CERN in Geneva (left) has a circumference of 27km. The wide-bore tube to the right of the operator is a magnet designed to narrow the beam of charged particles in the accelerator.

POWELL AND THE PION

The Japanese physicist Hideki Yukawa was the first to predict, in 1935, the existence of mesons – carrier particles of the strong nuclear force, one of the four fundamental interactions in nature. In an attempt to prove this hypothesis in 1947, before the advent of particle accelerators, British physicist Cecil Powell began investigating subatomic particles from space. As atmospheric cosmic rays collided with molecules of gas in the Earth's atmosphere, they created sprays of particles, as in an accelerator. Powell placed long-exposure silver-gelatin photographic emulsions on the tops of high mountains; when developed they revealed the tracks of charged subatomic particles, or 'pions' (short for 'pi mesons').

From the cyclotron to the synchrotron

The first cyclotron was built in 1931 at the University of California at Berkeley. It was a small machine, just a few centimetres in diameter, comprising two hollow metal half-cylinders placed side-by-side. Between them, an electromagnet created a perpendicular magnetic field, which bent particles injected into the cylinder into a circular orbit. Particle acceleration was achieved by two D-shaped electrodes set between the halves of the cyclotron, which generated an oscillating electric field that speeded up the particles every time they passed through the gap. Lawrence managed to accelerate protons to 1.25 megaelectron volts (MeV), as against Cockcroft and Walton's 0.7 MeV.

In 1945 Vladimir Veksler in the USSR, Edwin McMillan in the United States and Mark Oliphant in Britain all independently devised the synchrocyclotron. In this, the frequency of the driving electric field was varied with each increase in proton velocity, timed to synchronise with the passage of the particle beam, an extremely precise operation as the protons travelled at almost 300,000km per second. These machines could accelerate ions up to an energy of 400 MeV.

The golden age of accelerators

One further problem remained to be solved: during acceleration, particles tended to move away from the centre of the accelerator towards its walls. Yet the path described by the particles was governed by the magnetic field. To prevent particles from exiting their orbit, it would be necessary constantly to adjust

Pioneering accelerator *The Cosmotron at Brookhaven National Laboratory, New York. Protons were accelerated first in a Van de Graaff generator (the large drum, left foreground) before being fed into the circular cosmotron (behind, with a missing segment being installed).*

the strength of the magnetic field. This spurred scientists to create a more sophisticated form of particle accelerator – the synchrotron.

In 1947 construction of two synchrotrons began in Brookhaven, New York, and in Berkeley, California. The former had a radius of 9 metres, the latter 17 metres. In 1952 the Brookhaven synchrotron broke the billion electrovolt barrier, called a gigaelectron volt (GeV). Seven years later the European Organisation for Nuclear Research (CERN) began operating a 24 GeV accelerator, the Proton Synchrotron, at Geneva in Switzerland. Smashing particles into one another with this amount of energy gradually enabled physicists to discover all the elementary particles of which matter is made up.

The magnetic field remains the major obstacle preventing an increase in the power of synchrotrons. The greater the velocity of particles, the stronger the field must be in order to bend their trajectory. The Large Hadron Collider (LHC) is fitted with magnets that generate fields in the order of 8.3 Teslas,

The Large Hadron Collider *The huge particle accelerator near Geneva is buried underground at an average depth of 100m. The protons that are generated* (1) *are accelerated through the linear accelerator, or 'linac'* (2), *the tori of the Booster* (3), *the 628m circumference torus of the Proton Synchrotron* (4), *and finally through the torus of the Super Proton Synchrotron* (5), *which measures 7km in circumference. At this point, the protons are separated into two bundles of particles, which are sent in opposite directions* (6) *around two conjoined parallel tubes (coloured red and blue in the diagram) that make up the main 27km torus of the LHC. The particles collide where the tubes intersect* (7). *The first collisions were registered in late 2009 (inset) in the Compact Muon Solenoid. At 1,180 GeV, the LHC is already smashing all previous particle accelerator records, but this is nothing compared to its 7,000 GeV maximum energy.*

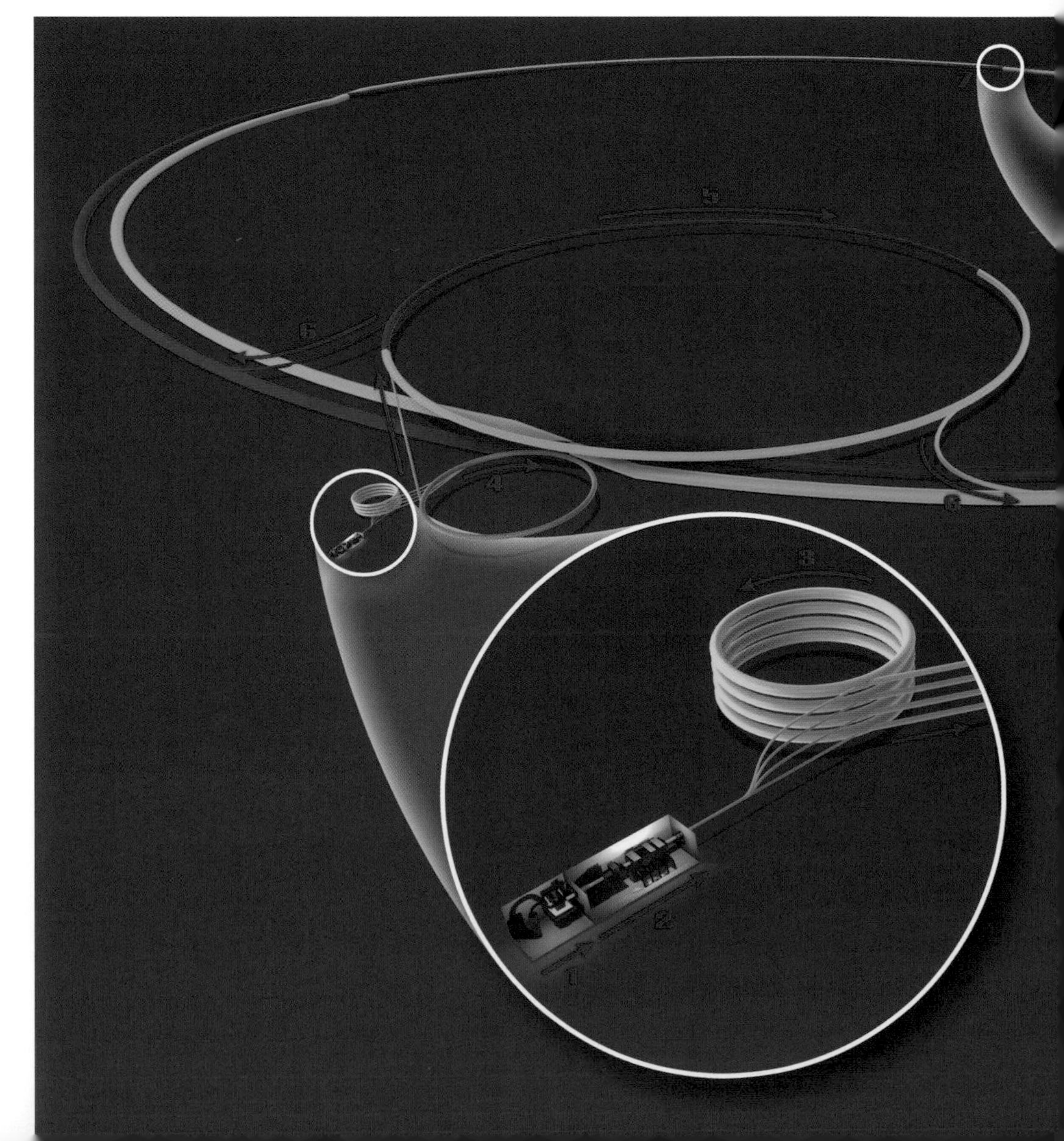

THE HIGGS BOSON

According to the standard model of particle physics, the Higgs boson is a hypothetical elementary particle that is thought to be responsible for gravity. In other words, the force of attraction that keeps the Moon in orbit around the Earth results from an exchange of Higgs bosons between the two bodies. The energy necessary to view this particle was hitherto beyond the scope of particle accelerators, but the LHC at CERN in Geneva is thought to have the capacity to detect it. If scientists fail to prove the existence of the Higgs boson, this will force a major rethink of many fundamental tenets of particle physics.

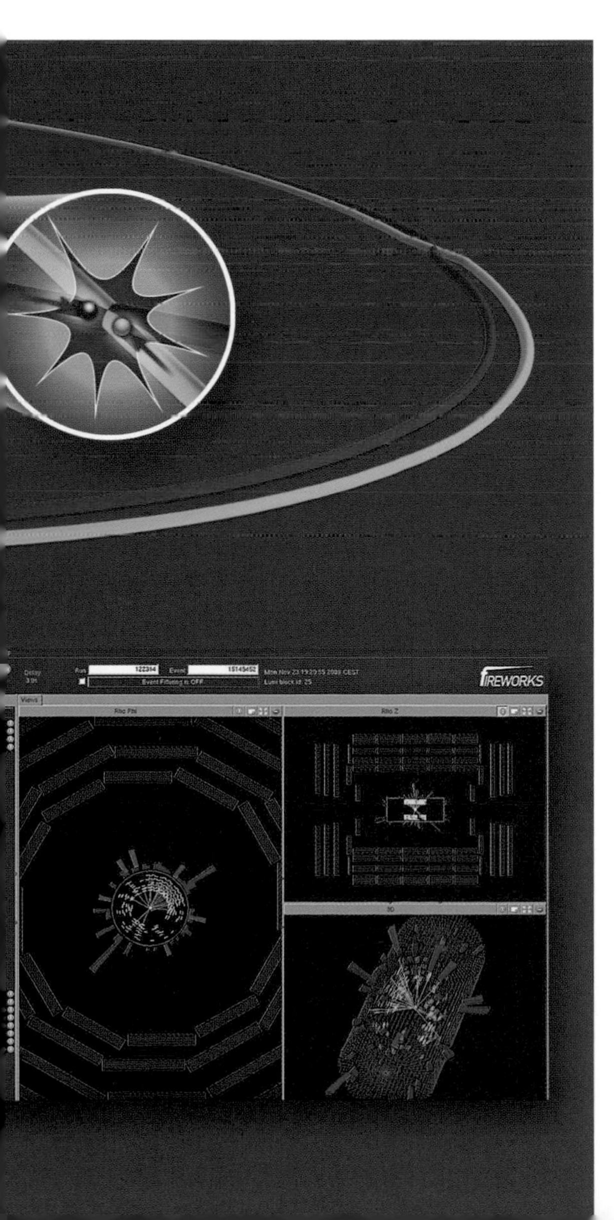

equivalent to 100,000 times the strength of the Earth's magnetic field. To achieve this, they must be cooled to –271°C, a temperature that allows the current to flow into the coils without melting them. By the time it reaches full energy, the hadrons (a type of proton) in the LHC will have reached 7,000 GeV, or 7 million times the energy of the first accelerators. The ultimate goal is to prove (or disprove) the existence of the Higgs boson, the particle that physicists believe gives mass to all matter in the universe.

Detecting collisions *The Compact Muon Solenoid (CMS) is one of several detectors installed on the Large Hadron Collider. Shown here is the silicon strip tracker, which identifies the tracks of individual particles.*

PARTICLE DETECTORS

The first piece of apparatus for detecting subatomic particles was the cloud chamber, invented by the Scottish physicist Charles Thomson Rees Wilson between 1897 and 1912. It was used to detect particles of ionizing radiation. The trajectory of a charged particle through a gas saturated with water or alcohol-vapour causes droplets of condensation to form, which leave behind distinctive tracks that can be photographed. In 1953 the cloud chamber was supplanted by the bubble chamber. The brainchild of American physicist Donald A Glaser, this reveals the tracks of subatomic particles as tiny bubbles in superheated liquid hydrogen. The entire chamber is surrounded by a strong magnetic field, causing the particles to deviate to a greater or lesser extent according to their mass and electrical charge. This allows each particle to be identified. The introduction of multiwire chambers by the naturalised French scientist Georges Charpak in 1968 marked the entry of particle detectors into the computer age. The trajectory of particles is shown by the electrical currents they generate while passing through a gas. These currents are detected by a fine mesh of wires within the chamber, and relayed to computers for analysis.

THE TRANSISTOR – 1947

Revolutionising electronics

In late December 1947, the physicists John Bardeen, Walter Brattain and William Shockley, who worked at the laboratories of the Bell Telephone Company, made history by creating the first point-contact transistor. They could scarcely have imagined the far-reaching technological revolution that their invention would set in motion.

Towards miniaturisation
A replica of the first working transistor invented at Bell Laboratories in 1947, which functioned like the 'cat's whisker' detector of earlier radio sets. The device was small by the standards of the day, about 10cm high, but miniaturisation in electronics has now progressed to the point where 7 million transistors can fit onto a single computer chip.

Breakthrough at Bell
The inventors of the transistor – Bardeen (right) and Brattain (left), with Shockley seated – at Bell Laboratories, the research division of AT&T.

Early radios, telephones and other electronic devices relied upon vacuum tubes to control their circuits: diodes, triodes and pentodes (with two, three and five electrodes respectively). In a vacuum tube, or 'thermionic valve', one heated electrode (the cathode or filament) emitted free electrons, which were attracted to the positively charged anode, forming an electric current. But valves had many drawbacks; not only were they difficult to manufacture, they were bulky and also extremely fragile. Early computers such as Colossus and ENIAC, which used thousands of such valves, were always breaking down.

The point-contact transistor

The first transistor made use of the curious properties of certain substances – germanium, tin, zinc and most famously silicon – known as semiconductors. A semiconductor is sometimes an insulator but under certain conditions it allows an electric current to flow through it. The researchers at Bell Laboratories used germanium, but had discovered that impurities in this element improved its semiconducting properties. This process, known as 'doping', enabled their small device to perform the three fundamental electronic functions – namely, to switch electronic signals on, to switch them off and to amplify them.

By attaching two gold point-contacts onto the germanium crystal, just fractions of a millimetre apart, Bardeen and Brattain achieved an output power far greater than the input. Thus was born the world's first solid-state amplifier. Unlike valve amplifiers, this one did not need to warm up before it could work. They presented their device – the world's first point contact transistor – to William Shockley, the leader of Bell's Solid

Ever tinier
A close-up of a revolutionary new bipolar junction transistor in February 1952. At just 15mm high, it was the world's smallest electronic component.

State Physics Group, on 23 December, 1947. He immediately recognised its potential, but it was six months before they went public in an article in the *New York Times* on 20 June, 1948, entitled 'The News of Radio'.

The junction transistor

Shockley now played a key role in the development of the first viable mass-produced transistor, the bipolar junction transistor (BJT). Building not only on Brattain and Bardeen's invention, but also on work on the purification of germanium by Morgan Sparks and Gordon Teal, two more colleagues at Bell, in 1951 Shockley came up with the bipolar transistor consisting of three semiconducting terminals: the collector, the base and the emitter. The

MISSED BY A WHISKER

As Bardeen, Brattain and Shockley were announcing their invention, two German physicists, Herbert Mataré and Heinrich Welker, were conducting independent research in the same field. While working at a subsidiary of the US firm Westinghouse in Paris, on a contract for the French Postal and Telecommunications service, they invented the 'Transistron', a point-contact germanium transistor very similar to the one developed at Bell. Unveiled on 18 May, 1949, it was hailed by the French government as a 'brilliant triumph of French research' and patented in August 1948, but no attempt was made to contest the American transistor's precedence. American teams sustained their innovation in transistors at Bell Labs and other companies such as Texas Instruments (which created a UK headquarters in Bedford in 1957), leading to their early domination of the field before competition arrived from Japan.

Quantum leap
The JOHNNIAC computer, conceived by the mathematician John von Neumann in 1953, used thermionic valves as its active components. Several years later, these were replaced by far less bulky and more reliable transistors (inset). The two pictures above are on the same scale.

collector was where the electricity supply was stored, the base was what controlled the 'gate' holding back this large electricity supply; the emitter was the outlet for the supply.

Far simpler to manufacture than the point-contact transistor, the new BJT became an indispensable component in all electronic devices (and would remain so until 1970 when an even smaller component that used less power was introduced – MOSFET, the metal-oxide semiconductor field-effect transistor). Although initially electrical engineers remained wedded to the familiar vacuum tube, they were gradually won over by the bipolar junction transistor's clear advantages: it was far smaller, easier to make, required far less power and was more reliable.

For their achievements Bardeen, Brattain and Schockley were jointly awarded the Nobel prize for physics in 1956. Bardeen subsequently left Bell Laboratories and in 1972 became the only man to become a two-time Nobel physics laureate for his work with Leon Cooper and John Schrieffer on superconductivity.

The dawn of modern electronics

The transistor's low electricity consumption made it ideal for use in a whole range of electrical equipment. The device in which it had the clearest impact was the transistor radio, which freed thousands of listeners from heavy sets running on mains electricity. The first battery-powered transistor radio went on sale on 18 October, 1954 – the TR-1, a joint venture by Texas Instruments and the engineer Richard C Koch of IDEA (Industrial

BIPOLAR AND UNIPOLAR

The bipolar junction transistor (BJT) – the one shown below is from the early 1960s – is so called because it consists of two junctions between three layers of semiconducting material. Depending on the impurities added to the basic element, two different types of semiconductor can be created: the n-type, with free electrons carrying a negative charge, and the p-type, in which 'vacancies' (or 'holes') carry a positive charge. The chemical bonding behind this process works as follows: if arsenic (with five outer electrons) is added to germanium, each arsenic atom links to four silicon atoms, leaving one surplus electron to carry a negative charge. Conversely, if germanium is doped with indium (which has three outer electrons), all available electrons are used, leaving a vacancy for a positively charged proton. Junction transistors can have either a p-n-p or an n-p-n structure. In the former, the emitter and collector consist of p-type material, while the base is a thin layer of n-type – and vice-versa in the latter. Depending on how it is connected into a circuit, the junction transistor can act as a current or voltage amplifier, in much the same way as a triode valve.

The principle of the field-effect transistor (FET) was outlined in a patent by the German physicist J E Lilienfeld in 1925, but the lack of suitable semiconductor materials meant that the first practical FET did not appear until 1952. The FET is a unipolar device made with just one type of doping, in which current is carried by one type of charge. The current flows parallel to the junction between the emitter (source electrode) and the collector (drain electrode).

Development Engineering Associates). Koch devised the circuitry (four germanium transistors and two diodes) while the firm Regency Electronics, a subsidiary of IDEA, was responsible for its manufacture.

The TR-1, the first 'shirt-pocket transistor' was just 13cm tall, ran on a 22.5-volt battery providing over 20 hours of life and retailed for around 50 US dollars. Other manufacturers soon followed suit, and the 'tranny' soon became a hit with consumers.

Shockley later followed in Bardeen's footsteps and quit Bell Labs. He founded the Shockley Semiconductor Laboratory at Mountain View in California, one of the first companies to set up shop in the area that would later become famous as 'Silicon Valley'.

WHY 'TRANSISTOR'?

The engineers at Bell Labs racked their brains to come up with a snappy, memorable name for their revolutionary invention and help to win over the sceptics. They asked for suggestions from their colleagues and one term, suggested by John Robinson Pierce, emerged as the clear winner, being adopted officially on 28 May, 1948. As Pierce later recalled: 'The way I provided the name was to think of what the device did. And at that time, it was supposed to be the dual of the vacuum tube. The vacuum tube had transconductance, so the transistor would have "transresistance". And the name should fit in with the names of other devices, such as varistor and thermistor. And so … I suggested the name "transistor".'

Portable and stylish
In the 1960s transistors made heavy old radio sets a thing of the past. Advertising, as in this Philips poster, emphasised the lightness and portability of the new radio, and the freedom from a fixed power supply.

News at your fingertips
The transistor allowed people to keep up with news and current events wherever they were. The anxious listeners below are Czech citizens listening to news about the Soviet invasion on 29 August, 1968.

CLAUDE LÉVI-STRAUSS – 1908 TO 2009

The father of modern anthropology

Claude Lévi-Strauss, who died in November 2009 at the age of 100, was the author of some 20 major works on anthropology, including *A World on the Wane* (1955), *The Savage Mind* (1962) and *The Raw and the Cooked* (1964). His radical holistic approach spread far beyond the social sciences, influencing disciplines as diverse as literary criticism and music.

Among Amerindians *A photograph taken by Lévi-Strauss of the funerary rites conducted by the Ewaguddu clan of the Bororo people in the Mato Grosso region of Brazil (below). The masked dancers are clad head to toe in foliage and are wearing feathered diadems. The brightly coloured headdress (right), incorporating parrot and macaw feathers, was brought back by Lévi-Strauss from Mato Grosso.*

Born in Brussels to French-Jewish parents, Claude Lévi-Strauss is renowned for having applied to his chosen field the analytical method known as structuralism. Its unique approach was to study the whole gamut of social phenomena – such as family relationships (especially the taboo on incest), myths and artistic endeavour – as individual elements of an organised system and not as arbitrary or irrational rules.

Structuralism shed light on unconscious thought patterns common to all humanity. It posited that all societies, irrespective of where they lived, were governed by identical structures, however much the superficial aspects of these laws and norms might appear to differ from region to region. Lévi-Strauss encapsulated this idea in the concept of the 'savage mind', which represented a radical departure from the traditional attitude to supposedly 'primitive' cultures in Western societies convinced of their scientific and technological superiority.

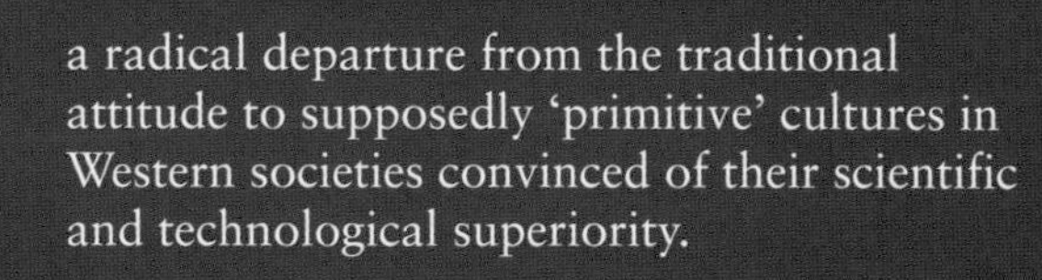

An itinerant scholar

After graduating in philosophy in 1931, Lévi-Strauss went to Brazil in 1935 to become professor of sociology at the recently founded University of São Paulo. Over the next four years he gained a thorough grounding in ethnography through a series of field trips to the remote Mato Grosso and Amazonia regions, where he lived among Amerindian peoples such as the Kaingang, Caduveo, Bororo, Nambikwara and Tupikawahib.

Out in the field
Claude Lévi-Strauss, photographed in Amazonia in 1936. The anthropologist spent a total of three years in the field and his work was hugely influential, though critics accused him of making his findings fit his preconceived theories.

Seminal insight
In his 1955 work Tristes tropiques, *published in English as* A World on the Wane, *Lévi-Strauss argued that both the 'savage' and 'civilised' mind are characterised by the unique and universal ability of human beings to see an object not merely as itself, but also as a symbol.*

He returned to France on the eve of the Second World War, but after the German invasion found himself debarred from teaching by anti-Semitic laws. In March 1941, aided by the Rockefeller Foundation's programme to save European intellectuals from Nazi persecution, he managed to emigrate to the United States and took up a visiting professorship at the New School for Social Research in New York. Fellow émigrés and friends included the Surrealist artists André Breton, Max Ernst and Marcel Duchamp, the anthropologist Franz Boas and the linguist Roman Jakobson, whose mathematical theory of language made a lasting impression on him.

A universal message

On his return to France in 1948, he submitted his thesis on 'The Elementary Structures of Kinship', but it was his 1955 account of his pre-war travels in Brazil *Tristes Tropiques*, translated into English as *A World on the Wane*, that made his name. The 1960s and early 1970s then saw publication of his major four-part work *Les Mythologiques*. These studies displayed the huge scope of his interests, including cookery, music, literature and Eastern religions. As his influence grew, so structuralism came to be applied to other areas of the social sciences and then to other disciplines.

In his analysis of man's place in the natural world, his belief in shared 'universal attributes' and his critique of the way in which 'progress' was creating a homogenised world, Lévi-Strauss anticipated much of the current debate about the future of the planet.

A HOMAGE TO FIRST-NATION PEOPLES

Lévi-Strauss began *A World on the Wane* with the words: 'I hate travels and explorers. Yet here I am proposing to tell the story of my expeditions.' But he did more than that in his work – he cast doubts on the supposed 'benefits' of Western civilisation. The time he spent with the Amerindians of Brazil made him reassess his attitude to the 'civilised' world, which seemed to him to export war and destruction, hastening the extinction of aboriginal peoples and the degradation of the ecosystem. He delivered a scathing attack on Western values, stating: 'The first thing we see as we travel round the world is our own filth, thrown into the face of mankind.'

HOLOGRAPHY – 1947

Images in the round

While trying to improve the resolution of the electron microscope, the Hungarian-Jewish émigré Dennis Gabor, who was working for the Thomson-Houston company in Rugby, developed the basic concept of holography – the reconstruction of an image in three dimensions. But his idea had to await the invention of lasers to become a reality.

Setting up
Positioning a laser for a hologram recording (below). From 2009, thanks to the invention of a portable device by the French laser physicist Yves Gentet, holograms can be produced outside specially equipped studios.

Holography is based on the principle that light is made up of a series of waves. If you throw a stone into a pond, the initial splash creates concentric circles of small ripples. Each point on the surface forms a crest that then becomes a trough until the motion ceases. At any moment, that point can be defined in terms of its height and its precise position in the cycle of rising and falling. These two parameters are known as the amplitude and the phase. In the same way, every point in a beam of light can be characterised by its luminous intensity and its phase. A photographic plate captures the intensity of light at each point, resulting in a two-dimensional image. But to re-create depth in the visual field, it is necessary also to capture the phase. Yet no physical device – nor indeed the human retina, in which the depth perception relies on parallax – is sensitive to this. Dennis Gabor therefore conceived the idea of capturing it indirectly.

To return to the pond analogy, if two stones are thrown into the pond simultaneously, a patch of smooth water forms in the zone where the two series of waves meet. At certain points where the phases of the waves are opposed to one another, the waves cancel each other out and their motion ceases, while at other points where their phases are the same, the two waves combine and reach their maximum height. This two-stone pattern therefore contains information both about the amplitude and the phase.

Gabor's technique was to bring two beams of light to bear on a photosensitive plate. The first, the so-called 'reference beam', was aimed directly at the plate, while the second – the 'object beam', which resulted from light being reflected off the target – arrived indirectly. The plate captured the pattern arising from this interference between the beams. The seemingly random pattern bore no resemblance to the object, but if the reference beam was then shone on the plate once more, the light field resolved itself into a three-dimensional image.

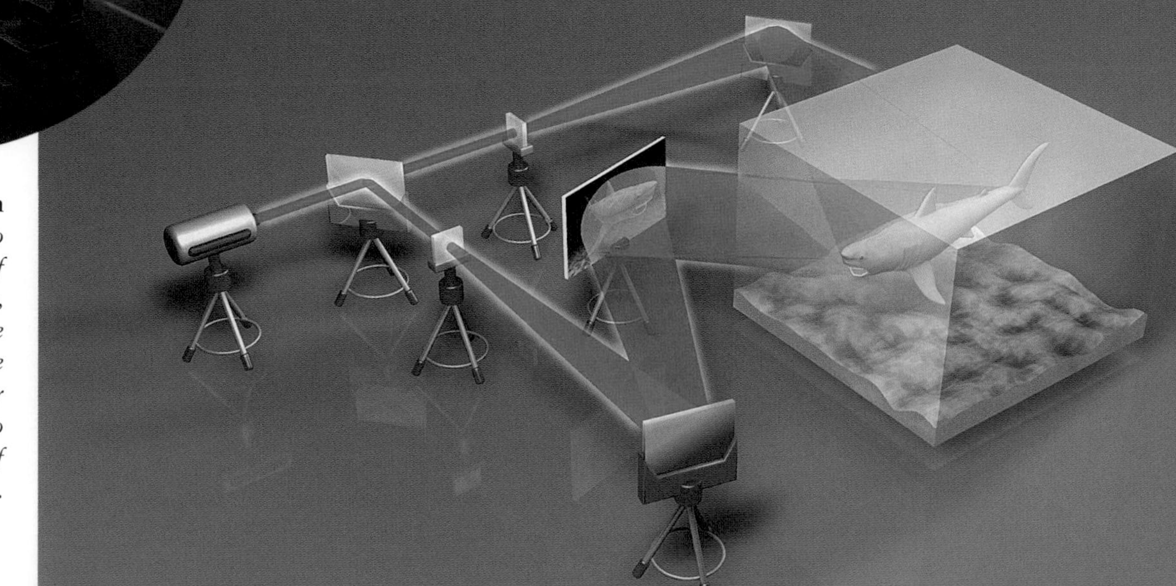

Creating a hologram
The laser beam is split into two separate beams. One of these illuminates the object, reflecting a 2D image onto the photographic plate (in the centre); the other beam is projected onto the plate via a series of intensifying mirrors.

Spectral apparition *Through advances in digital technology, ghost-like holograms can be incorporated into live-action cinema, as in this image of Jedi Council member Ki-Adi-Mundi in a scene from* Star Wars III: Revenge of the Sith.

THE EXPLODING WORLD

In March 1984, *National Geographic* magazine became the first major publication to put a rainbow transmission hologram on its cover. Four years later, on its centennial edition in December 1988, the entire cover was a hologram. The image chosen was of the Earth, and when turned in a different light it appeared to explode into fragments, symbolising the fragile state of the planet. This second image was captured by creating a hologram – produced with a series of exposures of just a billionth of a second – of a glass globe shattering under the impact of a bullet.

A bright idea ahead of its time

In his acceptance speech on receiving the Nobel prize for physics in 1971, Gabor commented ruefully: 'We had started 20 years too early.' Obtaining an interference pattern required a completely coherent light source – one that would generate photon waves with a definite phase, that is, a single, consistent wavelength. This was only achievable through the use of the laser, which was not invented until the early 1960s. As a result, the first practical optical holograms were produced in 1963 by Yuri Denisyuk in the USSR and by Emmett Keith and Juris Upatnieks in the USA.

In 1968 the American physicist Stephen Benton of the Polaroid Corporation devised a hologram that could be viewed with ordinary white light, without the need for a laser. It was named rainbow transmission hologram from the effect it produces of a range of colours.

Artists and counterfeiters

For a long time, lasers appeared to have little practical purpose and remained something of a sideshow confined to science museum exhibits and art installations – Salvador Dali, for instance, was interested in them. Then in 1974 a method was devised for stamping holograms into plastic on an industrial scale at minimal cost. Soon, holograms – which are impossible to fake without the original image – began to appear as anti-fraud 'watermarks' on bank cards, passports, tickets and other items susceptible to forgery.

Industrial holography

Holography is now widely used in industry as a quality-control technique. For example, the hologram of a product at any given stage in the production process can be compared to one taken earlier to identify and correct any deviations. Such applications took off with the digital age: the interference pattern could be captured by digital camera and instantly turned into an image on a computer screen. The American researchers Joseph Goodman and R Lawrence were the first to raise the possibility of digital image formation from holograms in 1967, while the first actual digital hologram appeared five years later in the USSR. The technique evolved in the 1990s, opening up new potential applications such as holographic films in real time or data storage.

THE WHOLE MESSAGE

The term hologram was coined by Dennis Gabor from the Greek words *holos*, meaning 'whole,' and *gramma*, meaning 'message'.

Card security *Holograms on bank cards are known as 'rainbow transmission holograms'. They need no special light source to appear, so can be seen in normal daylight.*

The autopilot 1947

On 22 September, 1947, a Douglas C-54 Skymaster of the US Air Force, a four-engined transport plane with a 35-metre wingspan, took off from Nova Scotia and flew some 2,500 miles to land in Britain without any of the 14 crew members touching the controls. This remarkable achievement was made possible by the invention, back in 1852, of the gyroscope. Although a rotating device called Serson's Speculum had been invented in 1743, the key model for the autopilot was the brainchild of French physicist Léon Foucault: a spinning wheel whose axle remained stable whatever the motion of the platform it was mounted on. The automatic pilot on the C-54 had a system of gyroscopes that permanently corrected the aircraft's controls (via electric servo motors), thus maintaining its set altitude and course and keeping it straight and level.

In 1912 the American engineer Elmer Ambrose Sperry fitted the first prototype gyroscopic stabiliser to the controls of an aeroplane. It was demonstrated by his son Lawrence at the Air Safety Competition held in Paris in June 1914. He released the controls and instructed his mechanic to walk out on one wing to destabilise the aircraft. In an instant, the stabiliser caused the plane to correct its attitude.

Following the success of the 1947 flight, the autopilot became a standard feature on most commercial and military aircraft. Constantly improved in the ensuing years, it has made a major contribution to air safety.

Straight and level
The instrument known as the artificial horizon shows the pilot the aircraft's attitude in relation to the horizon line. If need be, the autopilot adjusts the plane's attitude by operating the controls.

The Frisbee 1947

A game for all
The Frisbee's cheapness and simplicity made the game an instant hit on student campuses across America and in myriad locations worldwide.

In the 1920s a new craze began among students at Yale University in New Haven, Connecticut. They started throwing and catching empty pie tins, rather like the discus throwers of ancient Greece. The Ivy Leaguers got their tins from a local baker in Bridgeport called the Frisbie Pie Company.

In 1947 Walter Frederick Morrison, a recently demobbed US Air Force pilot, and his business partner Warren Franscioni hit upon the idea of a ready-made flying disc toy made out of Bakelite. At the time, the country was gripped by stories of alien spacecraft – notably the Roswell Incident, the supposed crash of an alien craft in New Mexico in July 1947 – so they marketed their disc as the 'Flyin' Saucer'. In 1951 Morrison launched the 'Pluto Platter', an improved version that set the definitive shape. In 1955 he and Franscioni sold the rights to the Wham-O toy company (later responsible for the Hula Hoop). Renamed the Frisbee, after the well-known Yale game, it was launched in 1957 and took the world by storm over the following decades, becoming especially popular as a beach game. The first official set of rules was drawn up in 1967.

ULTIMATE FRISBEE

This game, also called 'Ultimate', was invented at Columbia High School, New Jersey, in 1968. It involves a 175-gram disc – slightly heavier than a normal Frisbee – and points are scored by flipping the disc to a player to catch in the opposing end-zone (as in American Football). Each side has seven players, and players may move only one foot while holding the disc.

The Polaroid camera 1948

Edwin Herbert Land was just 17 years old when he won a place at Harvard in 1926. He enrolled to take a degree in chemistry, but left after his freshman year. Land's temperament was more entrepreneurial than academic, and he was keen to exploit his own scientific inventions and ideas as soon as possible.

Moving to New York, in the late 1920s he invented the first inexpensive transparent filters that could polarise light. These involved aligning microscopic, needle-shaped crystals of a polarising substance and sealing them inside a sheet of plastic. He called this process 'Polaroid', which also became the name of the company he founded in 1937. Before and during the Second World War, the Polaroid Corporation became known for its sunglasses and photographic filters.

Magic moments
On 27 October, 1972, Life magazine devoted its cover to Edwin Land and his revolutionary new SX-70 camera. It folded down into a slim, compact case that could fit easily into a suit jacket pocket.

POLAROIDS AS ART

The 'Polaroid Movement' arose among Pop artists of the 1960s, such as Andy Warhol and David Hockney, who valued the immediacy and repetitiveness of Polaroid images. In the 1980s artists found that they could manipulate the soft, slow-hardening emulsion on SX-70 film to startling effect: the satirical artist Ralph Steadman used the technique to create grotesque 'Paranoids', as he called them, of politicians and celebrities.

A new way of taking photos

In 1947, Land devised a revolutionary new 'instant photo' system, in the form of a camera that could develop its own pictures on special film. All the user had to do was click the shutter then wait about a minute before opening up the camera back and lifting out a black and white photograph. All of the processing, including exposure, development and fixing of the image, was done automatically inside the camera body. Launched in 1948, the Polaroid 95 was an instant hit. In 1962, Land adapted his system to colour by creating Polacolor film. Ten years later, he introduced a new generation of lighter, more versatile cameras, the SX-70 series. The camera automatically ejected the picture on shutter release, which then developed in a few minutes.

Despite major investment, Polaroid sales were hit badly by the spread of one-hour photo processing labs and then killed off by the advent of digital photography. The company went into receivership in 2008.

Instant exposure
The novelty of instant photos made the Polaroid camera an international success story. Even Tibetan monks in Lhasa (below) enjoyed a Polaroid snap.

THE BIG BANG THEORY – 1948

From a ‘primeval atom’ to a universe

The idea of the ‘Big Bang’ as the origin of the universe was first expressed in 1931 by the Belgian astrophysicist and Catholic priest Georges Lemaître. No-one took it very seriously until 1948, when the work of a number of scientists coalesced in two published articles. Thereafter the theory became firmly established as the prevailing explanation for the creation of the cosmos.

The current cosmological model, accepted by the vast majority among the scientific community, is that the universe was born some 13.7 billion years ago in what is commonly known as the ‘Big Bang’, a cataclysmic event releasing immense heat and ultradense energy. Since then, cosmic space has continued to expand, causing stars to become ever more distant from one another. The theory is so well known today, it is hard to imagine the uproar and bitter controversy that it gave rise to when the hypothesis was first aired in the 1930s.

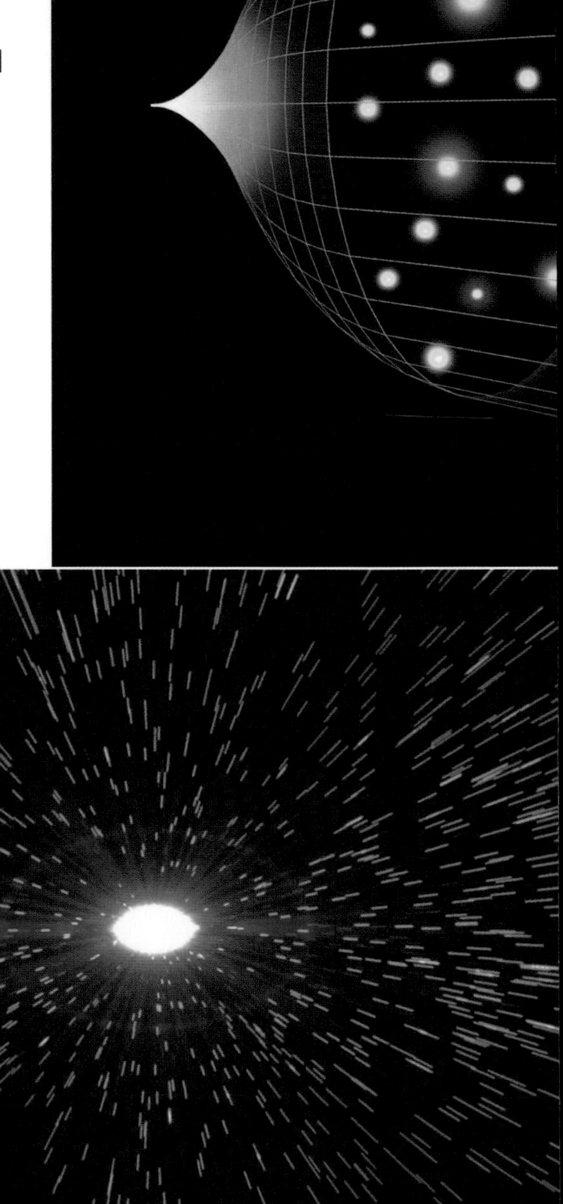

In the beginning ... *An artist's impression of the big event (below), some 13.7 billion years ago, when the 'primeval atom' exploded and the universe began.*

A dynamic cosmos?

Reflections on the origins of the universe are as old as astronomy itself. They took on a new dimension following the publication, by Albert Einstein, of an article entitled ‘Cosmological Considerations of the General Theory of Relativity’ in 1917. Einstein had recently promulgated his general theory of relativity, in which he concluded that the presence of mass or energy will curve space, slowing down time, in the immediate vicinity. He then began to apply his equations to the entire cosmos.

If a mass bends space-time locally, then what effect will all those curvatures have on the universe as a whole? Working from a hypothesis which proposed, in its simplest terms, that matter is distributed evenly through space (the ‘cosmological principle’), Einstein arrived at the radical new insight that the total volume of the universe was in a state of flux, since it was constantly expanding and contracting. The inescapable conclusion to be drawn from his theory was that the universe – space itself – was a dynamic entity.

The static universe

As a classically trained scientist, Einstein struggled to reconcile himself to his own theory. He continued to maintain that the

OLDER THAN THOUGHT

In 1930 cosmologists estimated the age of the universe at around 1.8 billion years. Fifty years later, the consensus was that it was some 13.7 billion years old.

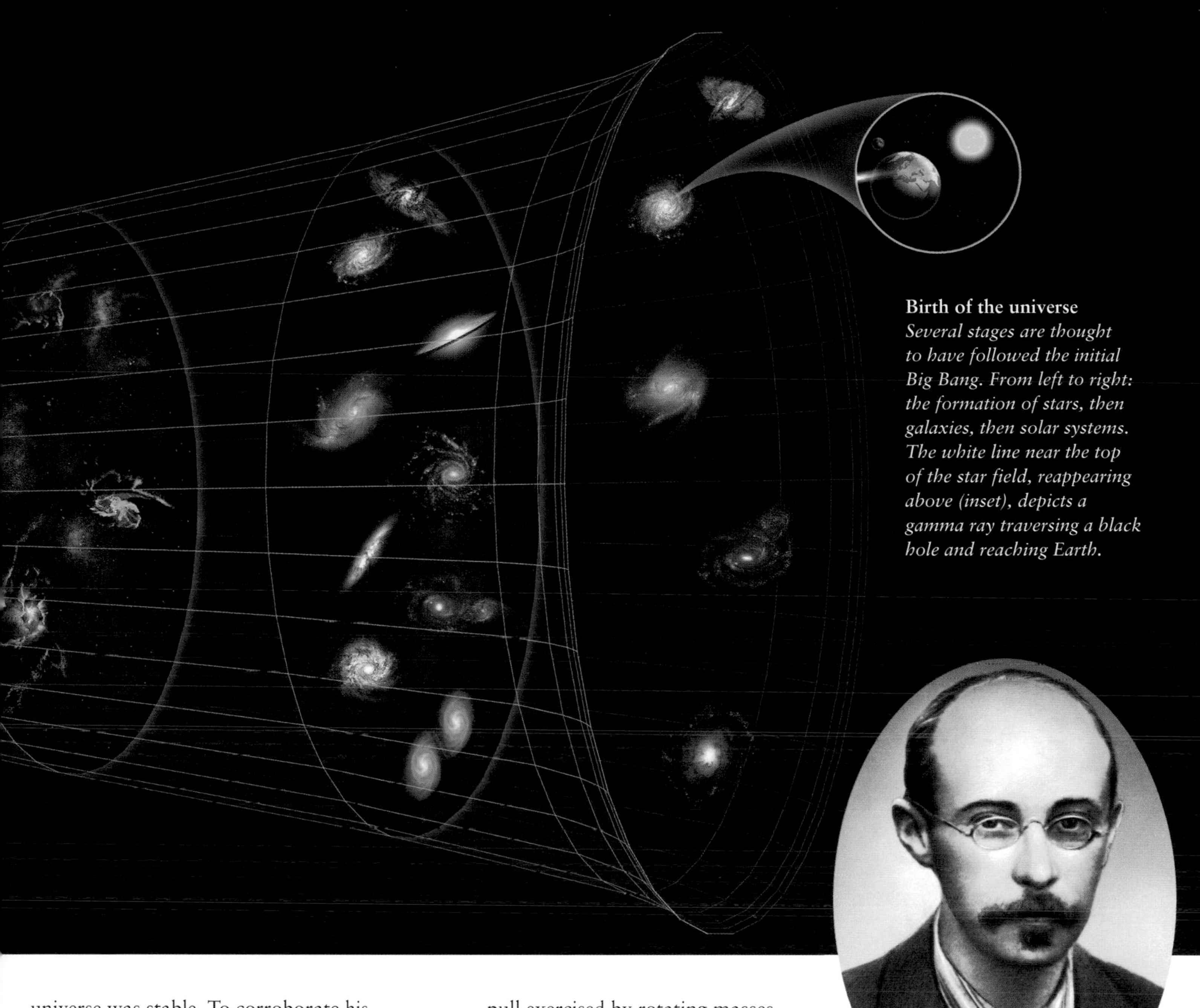

Birth of the universe
Several stages are thought to have followed the initial Big Bang. From left to right: the formation of stars, then galaxies, then solar systems. The white line near the top of the star field, reappearing above (inset), depicts a gamma ray traversing a black hole and reaching Earth.

universe was stable. To corroborate his standpoint, he resorted to subterfuge by introducing into his general equation the 'cosmological constant', a term designed to exclude any instability from the cosmos. This hypothetical force supposedly gave empty space the capacity to resist the gravitational pull exercised by rotating masses passing through it. According to Einstein, for the universe to remain static, there had to be a perfect balance between gravity and the cosmological constant. Einstein would later call his invention of the cosmological constant the 'biggest blunder of my life'.

Far from resolving the matter, Einstein's article only succeeded in reviving scientific interest in alternative models of a dynamic universe – expanding contracting, or a succession of the two. Between 1922 and 1924, the Russian mathematician Aleksandr Friedmann published his own model of the expanding universe. Confirming Einstein's field equations, he posited a big bang, followed by expansion, then gradual contraction, and eventually a 'big crunch'. Then in 1929 the American astronomer Edwin Hubble observed that galaxies appear to be moving away from one another, confirming that the universe was indeed expanding.

DITCHING THE 'BIG CRUNCH'

Cosmologists once thought the universe might eventually collapse under its own gravity into a black hole singularity, a point with zero volume and infinite density. The theory was known as the 'Big Crunch'. Yet the discovery in 1998 of a force called 'dark energy' suggested that expansion was in fact speeding up. Now physicists refer to the 'Big Rip', in which matter is progressively torn apart by the universe's expansion tens of billions of years in the future.

Father of modern cosmology
In 1922 Aleksandr Friedmann (1888–1925) corroborated the notion of an expanding universe and confirmed Einstein's General Theory of Relativity.

THE PRIEST-PHYSICIST

Georges Henri Lemaître was professor of astrophysics at the University of Louvain in Belgium, and also a Catholic priest. Although many of his scientific contemporaries were intrigued by his idea of the 'primeval atom', they were suspicious that he was trying to smuggle spirituality into science by the back door by postulating a kind of 'Genesis' moment. He was ultimately vindicated and in 1965, a year before his death, he had the satisfaction of seeing the discovery of cosmic microwave background radiation (CMBR) strongly corroborate the 'Big Bang' theory.

Ahead of his time *Georges Lemaître, architect of the Big Bang theory, in 1934, the year he was awarded Belgium's highest scientific distinction, the Prix Franqui.*

The theory of the 'primeval atom'

There now entered on the scene the true originator of the 'Big Bang' model of the universe, in the improbable form of a Belgian priest and professor of astrophysics, Monsignor Georges Lemaître. In 1931 he postulated the commonsense notion that, if the universe is expanding, it must have been much smaller in the past. Taking this idea to its logical conclusion, he claimed that the universe must have expanded from an initial point he called the 'primeval atom', prior to which nothing existed. Lemaître's ideas were largely ignored at the time.

The following year, 1932, Einstein and his Dutch colleague Willem de Sitter put forward a new model, which acknowledged expansion (already proved by Hubble) but suggested that this expansion is constantly slowing down. In other words, the universe was thought to be expanding indefinitely, but ever more slowly.

Gaining acceptance

In 1948 Lemaître's hypothesis and the Einstein–de Sitter model coalesced into a single theory: the Big Bang. This momentous event in the annals of science came about through the publication that year of two seminal papers: the first, 'The Origin of Chemical Elements', appeared in the April issue of *The Physical Review*; the second, 'The Evolution of the Universe', in the journal *Nature* in November. Not only were these the first studies to provide

Radiation layer *An image taken by the COBE satellite in 1992 (right) maps the cosmic microwave background radiation filling the universe. The deep blue areas mark the mean temperature of the radiation (2.73K), while the light blue and pink areas signify temperatures that are, respectively, 0.27K colder or warmer than the average.*

SUBSTANTIVE SUPPORT

Cosmic microwave background radiation (CMBR) was detected for the first time by Arno Penzias (above, on the right) and Robert Wilson (on the left), radio astronomers at the Bell Labs in New Jersey. While using a radiotelescope in an attempt to detect radio waves emitted by the Milky Way, they encountered a constant background noise that seemed to be coming from every direction. The average temperature of this radiation was close to absolute zero: 2.73 Kelvins or –270.45°C. This was in line with George Gamow's prediction of the existence of a trace of the Big Bang and was taken as corroboration of the theory. In 1992 and 2001, the COBE and WMAP satellites launched by NASA mapped the CMBR phenomenon, revealing microfluctuations around the mean temperature. The European Planck Space Observatory, launched in May 2009, will measure these fluctuations more accurately than ever before.

LIGHT YEARS

A light year is the distance covered in a year by a ray of light travelling through a vacuum at a speed of 186,000 miles per second (300,000km/sec). One light year equals 6 trillion miles (9.6 trillion km).

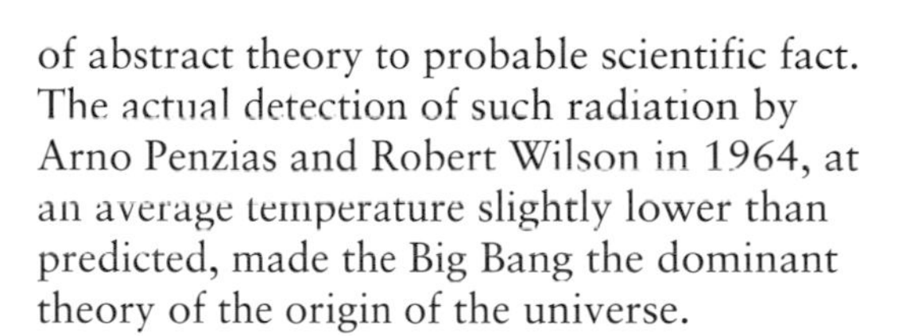

a concrete foundation for the Big Bang, they also posited the existence of an observable physical phenomenon that would prove the validity of the model. The Russian-American authors of the two papers – the first by George Gamow and his student Ralph Alpher, the second by Gamow alone – claimed that a trace of the original event must still exist in a wave of electromagnetic radiation filling the universe, testifying to the immense energy released by the Big Bang. They called this phenomenon 'cosmic microwave background radiation' (often abbreviated to CMBR) and estimated its temperature at somewhere between 5 and 50 degrees above absolute zero. The prediction of such an observable and quantifiable phenomenon immediately propelled the Big Bang theory from the realm of abstract theory to probable scientific fact. The actual detection of such radiation by Arno Penzias and Robert Wilson in 1964, at an average temperature slightly lower than predicted, made the Big Bang the dominant theory of the origin of the universe.

BEFORE THE BIG BANG

The Big Bang theory emerged from two major theories in physics – relativity and quantum physics – yet even these profound realms of inquiry cannot answer all questions. As a result, the model has potential flaws. For example, because quantum physics cannot probe smaller timescales than 10^{-43} seconds, Big Bang theory cannot tell us what occurred before this instant. In fact, our insight into what came before the first second in the life of the universe remains pure speculation. To confirm any surmise, we would need to artificially re-create energy densities several billion times beyond the capacity of even the most powerful existing particle accelerators, such as the Large Hadron Collider in Geneva. Beyond the pre-Big Bang period, particular stumbing blocks to our understanding are the phase transition into the so-called 'inflationary epoch' at 10^{-35} seconds, and the process by which antimatter disappeared.

Hyperinflation
This conceptual artwork (above) illustrates inflation in several areas of the early universe. The inflationary theory proposes that during the extreme conditions of the Big Bang, a 'false vacuum' created a repulsive force that caused an incredibly rapid expansion, one much faster than the ordinary expansion observed today.

GENESIS OF A UNIVERSE

The Big Bang began at t = 10^{-43} seconds (the Planck epoch), with an ultracompact universe that had between 10 and 25 spatial dimensions. At t = 10^{-36} seconds the phenomenon known as 'inflation' occurred: in an infinitesimal fraction of a second, the universe doubled in size a hundred times over, while most of its spatial dimensions collapsed into themselves, leaving just the three we know of today. Between 10^{-36} seconds and 10^{-6} seconds, as the universe continued to expand, cooling all the while, common particles began to form. These particles, called baryons, include photons, neutrinos, electrons and quarks – the building blocks of matter and life. Particles of matter were twinned with particles of antimatter: the two collided and destroyed one another until just a small amount of matter remained. From around t = 0.00001 seconds, at a heat of some 10 billion degrees, the first atomic nuclei were formed, of hydrogen and helium. This process continued until about 3 minutes after the Big Bang. Some 380,000 years after the event, when the temperature was at 4,000 kelvins (1K = 0°C + 273.15), came a stage called 'recombination', when hydrogen and helium atoms began to form. Over the following hundreds of millions of years, these combined to create stars and galaxies.

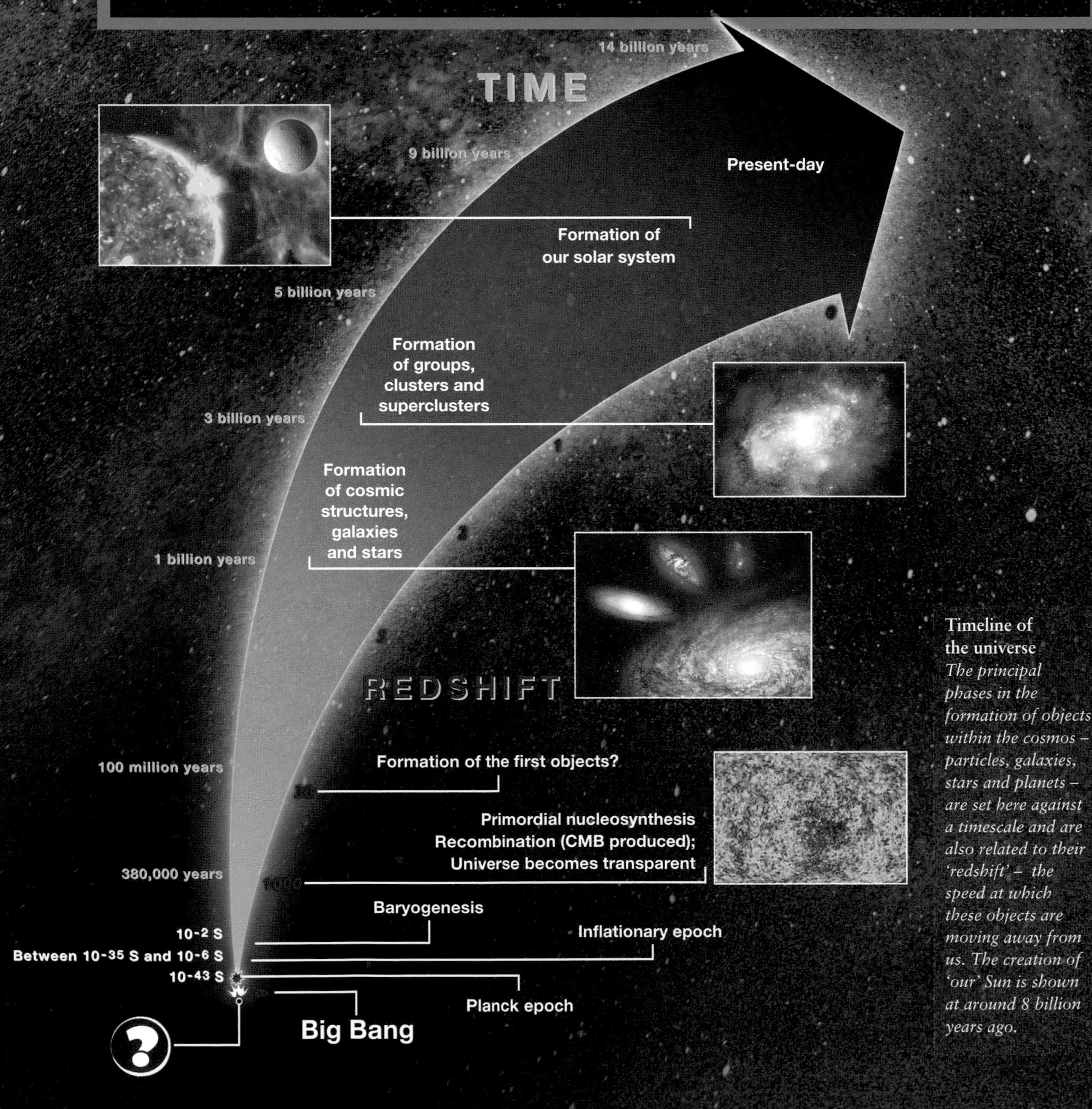

Timeline of the universe
The principal phases in the formation of objects within the cosmos – particles, galaxies, stars and planets – are set here against a timescale and are also related to their 'redshift' – the speed at which these objects are moving away from us. The creation of 'our' Sun is shown at around 8 billion years ago.

The Big Bounce
An alternative account claims that the first event in the history of our universe resulted from the collapse of a previous universe due to quantum gravity. The illustration above attempts to illustrate this 'big bounce'. The red spheres represent elementary particles of concentrated energy, separating from each other due to the expansion of the fabric of space-time.

ALTERNATIVE COSMOLOGIES

Although the Big Bang model remains the benchmark for cosmologists, its loopholes have prompted alternative explanations. In the 1950s, the British astronomer Fred Hoyle proposed a 'steady-state' cosmology in which the universe is subjected to short-term oscillations superimposed on a long-term steady expansion. In this model the universe has no beginning or end – new matter is spontaneously and continually created. A modification of this approach, the so-called 'Big Bounce' theory, conjectures that the Big Bang only occurred after an initial phase of contraction.

The universe on a pinhead

The term 'Big Bang' was coined in a 1949 radio broadcast by British astronomer Fred Hoyle who rejected the theory, arguing instead for a 'steady-state' model. Contrary to what the name suggests, the Big Bang was not an explosion in the true sense of the word, somehow creating the universe out of nothing. As currently understood, the model posits a pre-existing universe, which began expanding when it was just 10^{-43} seconds old, when space was in an extremely dense and hot state. If we can grasp the idea of a virtual sphere that is currently 13.7 light years in diameter (9.6 trillion km x 13.7), a few fractions of a second after the Big Bang this same sphere measured just 10^{-28} metres across; in other words, it was more or less a single point.

Yet this microscopic volume contained all the mass and energy that exists in the cosmos today, including gas clouds, galaxies, stars and so on, albeit in a different form. As this volume expanded and cooled, it led to the formation first of subatomic particles, then of atoms, molecules, stars, galaxies, planets and eventually life and the human race itself.

An evolving model

From 1948 onwards, the concept of an increasingly slowly expanding universe (Einstein–de Sitter), originating from a hyperdense and hot state (Lemaître), became the standard model used by cosmologists. Even so, it has been periodically challenged and modified as new data emerges. For instance, observations of certain supernovae in 1998 suggested that rather than slowing, the expansion of the universe has in fact been accelerating over the past 4 billion years.

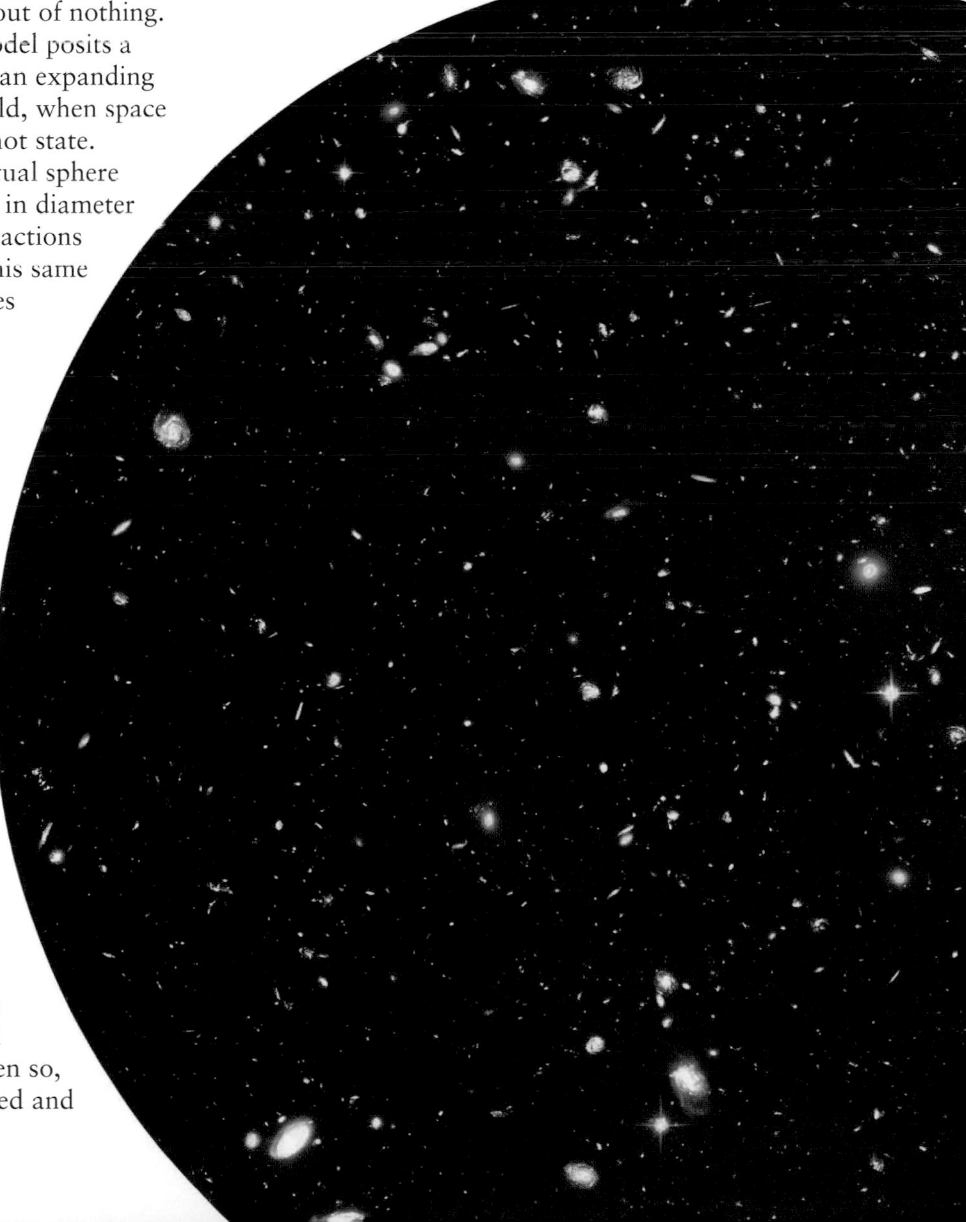

Going back in time
In 2004 the Hubble Space Telescope produced the deepest images of the universe yet taken. Photos like this one (right) show the first galaxies, which formed around 13 billion years ago when the universe was only 5 per cent of its present age – just after a period that cosmologists call the 'Dark Ages'.

GEORGE GAMOW – 1904 TO 1968

Physics for the common man

George Gamow, one of the architects of the Big Bang theory, combined an acute understanding of physics with an unconventional approach to science teaching and a flamboyant lifestyle. Endowed with a fertile mind and an irrepressible sense of humour, he was known for his love of dry martinis and pink Cadillacs.

George Gamow was born Georgy Antonovich Gamov at Odessa on the Black Sea coast in 1904, during the reign of Russia's last tsar, Nicholas II. For his 13th birthday, in the same year as the Bolsheviks came to power, the young Georgy, who was fascinated with star-gazing, was given his very first telescope. At university in Leningrad (St Petersburg), where he completed his doctorate in 1928, he briefly attended the lectures of mathematician Aleksandr Friedmann, the first scientist to advance the theory of an expanding universe. At that stage, Gamow's interest lay less in cosmology and more in the new discipline of quantum physics, which he studied first under Niels Bohr in Copenhagen and then in Cambridge with the radioactivity expert Ernest Rutherford.

Before coming to Western Europe, Gamow was known and widely respected for his theory of alpha radiation, which set out to explain why certain radioactive elements decay in just a few seconds, while others take thousands of years to do so. Working under Bohr, he devised his 'liquid-drop' model of the atomic nucleus, which was to prove extremely useful in understanding the mechanisms of nuclear fission and fusion. Yet Gamow never abandoned his youthful passion for astronomy, and his research into thermonuclear reactions within stars created a link between nuclear physics and astrophysics.

FLIGHT TO THE WEST

In 1931 Gamow was forced to leave Cambridge and return to Russia, but fearing that he might be sent to a Soviet labour camp in Siberia for his unorthodox views, he decided to leave his native country for good. Over the next two years he made several attempts to get out of the USSR, to no avail. His first attempt was a disastrous failure – together with his wife he tried to cross the Black Sea by kayak to Turkey, a voyage of some 170 miles, but bad weather forced them to turn back. In 1933 Gamow and his wife were granted permission to attend the Solvay Conference in Brussels, where they promptly defected to the West. He never returned to Russia.

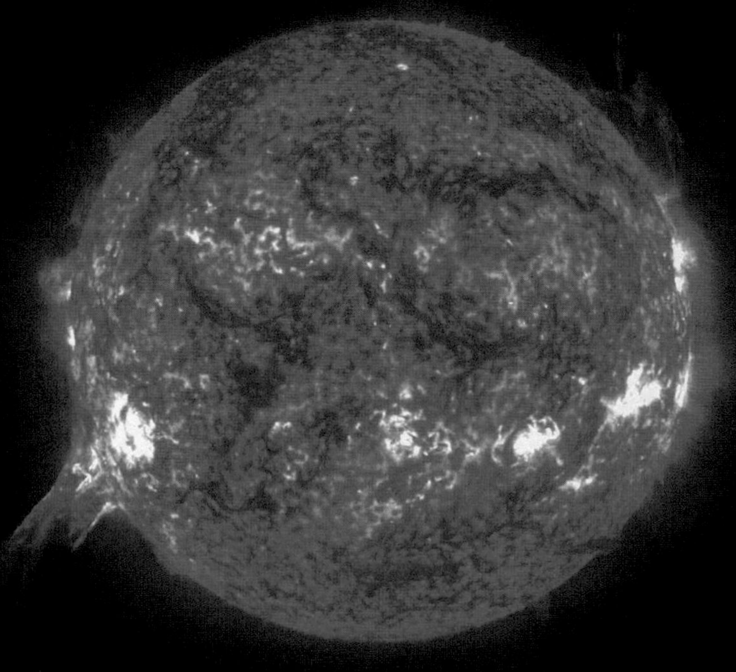

Thermonuclear fireball
Gamow was one of the first people to identify the massively powerful thermonuclear reactions that take place on the surface of the Sun. In 1995, the NASA probe SOHO sent back spectacular images, including this one (left), of the solar flares produced by these reactions.

Atoms and stars

In 1934 Gamow emigrated from Europe to the United States, where he secured a chair in physics at the University of Washington. Like thousands of other physicists of the era, he was coopted to work on the Manhattan Project to build the world's first atomic bomb. During this period, he continued to study the structure of the stars, publishing an important paper with Edward Teller on Red Giants in 1942.

After the war, together with his fellow Russian-American Ralph Alpher, he published a series of articles on the origins of the universe. Their 1948 work on the formation of the first elements, hydrogen and helium, showed how the preponderance of these in the universe – they make up more than 99 per cent of all matter – could be largely explained by the thermonuclear reaction that occurred during the 'Big Bang'. The atomic nuclei, they claimed, were produced by the successive capture of neutrons.

Gamow and Alpher, presently joined by the young Robert Herman, went on to hypothesise

Easy as one, two, three *George Gamow in his study in 1958. Two years earlier he had been awarded UNESCO's Kalinga Prize for his* 'Mr Tompkins' *series of books popularising science, published between 1939 and 1967. In 1999, more than 30 years after Gamow's death, the author Russell Stannard was asked to update the series to take account of recent scientific advances; the cover (above) shows the French edition of* The New World of Mr Tompkins.

that the formation of nuclei had generated high levels of radiation, traces of which must remain. In 1964–5 Penzias and Wilson detected the existence of this radiation, now known as cosmic microwave background radiation (CMBR), but then and for a long time after no acknowledgment was made of the original prediction. Gamow was incensed and fought a long campaign for his team's contribution to be recognised. Despite being universally acclaimed among his peers for providing one of the major proofs of the Big Bang theory, he was never awarded the Nobel prize for physics.

A great populariser

Perhaps Gamow's outgoing personality had something to do with him not always being taken seriously. He had a whimsical, mischievous sense of humour. He added the name of his friend Hans Bethe, an eminent Cornell physicist, to the list of authors of the 1948 paper just to create the pun Alpher, Bethe, Gamow (alpha, beta, gamma): Betha had nothing to do with it. Furthermore, the paper was due to be published on April 1st. Gamow's mind was a seething mass of ideas, which spilled out willy-nilly. He published a series of books, which he illustrated himself, based on a character he named Mr Tompkins, an everyman baffled by the theory of relativity and particle physics. The books made a major contribution in helping to explain modern scientific theories to a wider audience.

THE BATHYSCAPHE – 1948

Exploring the ocean deeps

Until the 1930s, the deepest parts of the world's oceans were completely unexplored. The depth record set in 1934 by William Beebe in his bathysphere remained a one-off achievement until Swiss explorer Auguste Piccard invented the bathyscaphe – a craft capable of diving into the deepest ocean trenches and surviving the massive pressure there.

The American naturalist William Beebe spent much of the 1920s on board ships, studying life forms trawled up from the sea floor. The specimens were often in a sorry state by the time he saw them – dead or dying, and badly damaged by the mesh of the trawl nets. Beebe decided he wanted to observe these fascinating creatures in their own habitat. With the help of engineer and fellow American Otis Barton, he had a special craft constructed. What they came up with was a sphere of cast iron, 1.34m in diameter with walls 3.75cm thick; it had one narrow entrance hatch and three portholes. The globular form meant that hydrostatic pressure was distributed equally around the hull of the craft. It could take two crewmen, either sitting or squatting in the cramped interior, and was tethered to the mother ship on the surface by a cable that commuted every movement from the surface – dangerous when the ship was rolling in a strong swell.

Between 1930 and 1934, Beebe made a total of 34 descents in his 'bathysphere' , reaching a depth of 1km (the actual record was 1,001 metres). Yet the shortcomings of this 3-tonne craft were evident. It could not move independently and if the cable broke, it would be a death trap.

An underwater balloon

While Beebe was making his descents, a Swiss physicist by the name of Auguste Piccard was heading in the opposite direction, becoming the first person ever to enter the stratosphere when, in 1932, his balloon *FNRS* attained a height of 16,201 metres. After the war Piccard transferred the principle he used in building the balloon to construction of a submersible: he suspended a very heavy crew gondola (a pressurised steel sphere) below a much larger 'float' that was filled with petrol (lighter than water). Contained within this

Record holder *A model of the bathyscaphe* Trieste, *launched in Naples in 1953. The craft was bought five years later by the US Navy.*

Cramped conditions *Otis Barton (left) on board the bathysphere he and William Beebe used in the 1930s. Their record-breaking dive took place off Bermuda in 1934. Beebe described the ocean floor as looking 'like the black pit-mouth of hell itself'.*

LIFE ON THE OCEAN FLOOR

It was long thought that life could not possibly exist in the deep trenches of the world's oceans, where it is permanently dark and the pressure is a crushing 1.25 tonnes per square centimetre. But in February 1977, when the American submersible *Alvin* reached a depth of 2.5km in the waters off the Galápagos Islands in the Pacific while investigating hydrothermal vents, the two geologists on board were astonished to find an abundance of life there. In particular, giant purple tubeworms thrived around the vents. Similar evidence of life has since been found in deep ocean waters elsewhere.

Extreme environment *Giant tubeworms (*Riftia pachyptila*) can grow up to 2m long at incredible depths.*

THE WRECK OF THE TITANIC

In 1912, the liner RMS *Titanic* sank beneath the Atlantic Ocean after hitting an iceberg off Newfoundland. Finding the wreck, which lay at a depth of 4km, became something of a holy grail for divers. It was located on 1 September, 1985, by a team led by the American Dr Bob Ballard and the Frenchman Jean-Louis Michel. As Michel reported: 'At 1.00 in the morning I spotted some faint images on the screens, which within a few minutes clearly turned out to be a ship's boiler. I had no doubt that this was the final resting place of the *Titanic*.' Over the next 12 years a manned submarine, the *Nautile*, dived repeatedly to the wreck, recovering numerous artefacts. In addition, a small ROV (remotely operated vehicle), *Robin*, took pictures of the stricken liner's interior.

float were two enormous hoppers filled with 4 tonnes of magnetised iron pellets, which acted as ballast. When diving, electromagnets held the ballast in place; to ascend, the pilot cut the electricity supply, releasing some of the pellets into the sea and thereby lightening the bathyscaphe. If anything went wrong with the electrical system when submerged, the ballast would automatically be released, allowing the craft to float to the surface. On 3 November, 1948, Piccard's bathyscaphe *FNRS-2* dived to 1,380 metres. Beebe's record had been smashed, but this was just a 'dry' run with nobody on board. Bad weather then prevented the craft returning to the surface and she was damaged while under tow back to port.

Deeper and deeper

Piccard's *FNRS-2* was bought by the French navy, who salvaged the crew gondola and incorporated it into her successor, the *FNRS-3*. Piccard and his son Jacques went on to build an improved version of the craft, which they named the *Trieste*. One key refinement was an entrance tunnel accessed by a hatch on top of the bathyscaphe, as on a submarine, through which the crew descended to the observation gondola. On 22 January, 1960, Jacques Piccard and American naval lieutenant Don Walsh descended to 10,916 metres, touching the bottom of Challenger Deep in the Mariana Trench in the Pacific. A new generation of lighter, more manoeuvrable submersibles have continued to explore the oceans since, studying wrecks, prospecting for oil and gas, collecting scientific samples. Yet their operational depth is limited by the relatively flimsy construction of the hulls, and the *Trieste*'s depth record stands to this day. Only two unmanned ROVs (remotely operated vehicles) – the Japanese *Kaiko* in the 1990s and the American *Nereus* in May 2009 – have even come close.

Live from the *Titanic*
In 1998 the Nautile *gave the world its first glimpse of the* Titanic *since the ill-fated liner sank in 1912. For two hours, the submersible sent back remarkable live pictures of what remained of the former pride of the White Star Line – this picture (inset left) shows the ship's bow railings – and broadcast them to an audience estimated at 250 million worldwide.*

AUGUSTE PICCARD – 1884 TO 1962

To boldly go ...

Auguste Piccard, inventor of the bathyscaphe, was also the first man to reach the stratosphere, closely followed by his twin brother Jean-Félix. The Piccard family's feats captured the world's imagination: Auguste was the inspiration for the eccentric Professor Calculus in the *Tintin* books.

Bold boffin
Auguste Piccard on the cover of a news magazine (above), just prior to his record-breaking ascent into the stratosphere in 1931. Piccard was the first person ever to see the curvature of the Earth.

Sibling rivals
Auguste's twin brother, Jean-Félix Piccard (above), in the steel pressurised capsule of the Century of Progress, *a balloon he designed and built for the first American ascent into the stratosphere. Piloting this craft, Jean-Félix broke his brother's altitude record in 1934.*

As a young man, the Belgian cartoonist Hergé used to see Piccard on the streets of Brussels – Piccard taught at the university there from 1922 – but never dared to approach him. Instead, he immortalised his childhood hero in the character of the absent-minded Professor Cuthbert Calculus. It was an affectionate caricature of the eminent scientist, and like all caricatures bore only a passing resemblance to the original. As Hergé himself recalled in an interview in 1948, Piccard was actually very tall: 'He had an interminable neck that sprouted from a collar that was much too large ... I made Calculus a mini-Piccard, otherwise I would have had to enlarge the frames of the cartoon strip.'

Precocious pranksters

Born in 1884 in Basle, Switzerland, Auguste Piccard had an identical twin brother, Jean-Félix. From an early age they were enthralled by the novels of Jules Verne, so their mother had a small laboratory set up in the family home where they could conduct experiments. The brothers were mischievous as well as inventive and loved to play practical jokes. On one occasion a barber, not knowing they were twins, was dumbfounded when one of them marched into his shop and demanded a free haircut, claiming the one he'd had the day before had grown back overnight.

Auguste and Jean-Félix had a thirst for knowledge and went on to study physics and chemistry respectively. Auguste's first papers were on magnetism and radioactivity. The physicist with round spectacles (with ingenious double frames of his own devising, enabling him to see both close-up and distance) secured teaching posts first at a college in Zurich and then at the University of Brussels. In 1920 he married Marianne Denis, a French student who had come to Zurich to improve her German. She and Auguste had five children – four girls and a boy. In his professional life, Piccard associated with the leading scientists of the time, including Albert Einstein, Max Planck, Niels Bohr and Marie Curie.

A hands-on scientist

Auguste Piccard was anything but an ivory-tower academic. Imbued with an unquenchable faith in scientific progress, he was always keen to get out and put his theories to the test. As he once put it: 'If the subject matter I'm studying is too big to fit into my laboratory, then I move my lab out into the subject matter.' To research the phenomenon of cosmic rays, he ventured into the stratosphere. To withstand the intense cold (–60°C), low air pressure and lack of oxygen at an altitude of 16,000 metres, he devised a balloon with a sealed pressurised cabin. He named the craft FNRS, short for Fonds National de la Recherche Scientifique, a scientific research fund set up by King Albert I of Belgium, which sponsored his work. On his first ascent, on 27 May, 1931, Piccard reached 15,780 metres. A second attempt, on 18 August the following year, soared to 16,201 metres. This record stood for two years, until it was broken by his brother Jean, who by then was teaching in the United States.

Auguste went on to set the world record for deep-diving in his bathyscaphe. He and his son Jacques then collaborated on improving the craft and demonstrating its practicality for undersea exploration and research. In 1962, at the age of 78, Auguste died of a heart attack. Jean-Félix never got over the loss of his brother and passed away a few months later.

AN EXTRAORDINARY FAMILY

The twin brothers Auguste and Jean-Félix Piccard founded a dynasty of intrepid scientist-adventurers. Auguste's son Jacques (1922–2008) was a renowned oceanographer who spent his life developing submersibles to explore the ocean's depths. In 1960, he dived deeper than any human being before or since. In 1999 Bertrand (born 1958) followed in his grandfather Auguste's footsteps by completing the world's first nonstop round-the-world balloon flight. The creator of *Star Trek*, Gene Roddenberry, honoured this illustrious family in *The Next Generation* by naming the admirable captain of the starship USS *Enterprise* (Galaxy Class 1701-D) Jean-Luc Picard.

Submariner extraordinaire *Jacques Piccard (above) in one of the submersibles he helped to develop after his record-breaking dive in the* Trieste *in 1960. The mesoscaphes (mid-level underwater craft) that he built included the first tourist submarine, the* Auguste Piccard, *in which he took passengers to the bottom of Lake Geneva during the Swiss Expo of 1964–5.*

Latest in the line *Bertrand Piccard (above) trained as a psychiatrist before continuing the family tradition of balloon exploration, flying non-stop around the world in 1999. He plans to repeat the feat in 2012 in a solar-powered aircraft, the* Solar Impulse.

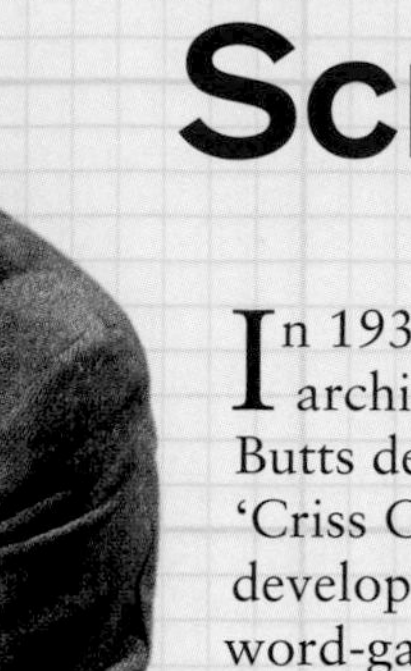

Scrabble 1948

In 1938 the American architect Alfred Mosher Butts devised a game he called 'Criss Cross Words', which he developed from an earlier word-game of his named Lexiko. Players had to make words from a random selection of letters, each of which yielded a set number of points (the least common letters secured the largest scores). Butts had carefully worked out the values allocated to each letter by analysing the frequency of use in various sources, including the *New York Times*. It was ten years before anyone took any commercial interest in the game. Then, in 1948, an entrepreneur named James Brunot purchased the rights to manufacture and market the game, paying Butts a royalty on every game sold. Apart from altering the scoring system slightly and simplifying some of the rules, Brunot retained the basic principles of the original game. He and Butts renamed it 'Scrabble'. By 2010 Scrabble was available in 29 different languages and was played in 121 countries. More than 150 million Scrabble games of have been sold worldwide.

Tough start
In the beginning, James Brunot and his wife made Scrabble *pieces by hand in the front room of their house in Newport, Connecticut, turning out 18 games a day.*

McDonald's® 1948

The first McDonald's® was opened as a Bar-B-Que restaurant in California by Richard and Maurice McDonald on 15 May, 1940, but it was eight years later that the foundations of the modern fast-food outlet were established. In that year, at their San Bernadino branch, the brothers introduced the 'Speedee Service System' and drive-in to maximise the turnover of hamburgers, which accounted for 80 per cent of their profits. In 1961 franchiser Ray Kroc bought out the McDonald brothers' stake in the company for $2.7 million. The McDonald's® Corporation he founded is now the largest fast-food chain in the world, serving more than 58 million customers daily.

Shrine to success
The ninth McDonald's® restaurant (below), opened in 1955 by franchiser Ray Kroc in Des Plaines, Illinois, is now the official museum of the McDonald's® Corporation.

DIET FOR DISASTER?

In 2004 the American journalist and filmmaker Morgan Spurlock created a stir with his documentary movie '*Super Size Me*', which recorded his experiment in eating only from McDonald's® restaurants for a month. During that time, Spurlock put on 11 kilograms in weight and suffered various health problems, including mood swings, sexual dysfunction and depression.

Barcodes 1949

After overhearing a conversation between a professor from the Drexel Institute of Technology (Philadelphia) and the owner of a local supermarket chain who wanted a tool that could instantly identify thousands of product lines in his shops, electrical engineer Bernard Silver persuaded his friend Norman Woodland to help him develop the barcode. Their first attempt consisted of lines traced in luminous ink that could be read under ultraviolet light, but the ink proved unstable and was too expensive to print on labels.

They persevered and after several months of research Woodland came up with a better idea inspired by a combination of the movie soundtrack synchronisation system, invented in 1920 by Lee DeForest, and Morse code technology. A bright light shone onto black and white bars produces a reflection of varying luminous intensity; picked up by a light-sensitive receiver and translated into digital code, this could be used to automate product data. On 20 October, 1949, Silver and Woodland lodged their first patent, comprising a bull's-eye pattern of black and white circles. But the system was still too complex and expensive to catch on. A decade would pass before simple and cost-effective barcode readers became a reality with the advent of the integrated circuit (1958) and the laser (1960). Thereafter, retailers quickly caught on to the new system.

The 1970s saw the adoption of the 12-digit universal product code (UPC), made up of a series of black and white vertical bars. This was the brainchild of an IBM engineer named George Laurer, who later added a 13th digit, thus creating the EAN-13 ('European Article Numbering') code, which became the international norm.

Self-scan service *Some supermarkets use a 'self-scan' system, in which customers scan each item as they shop, totting up the bill as they go. This greatly reduces the checkout time, with no need for goods to be unloaded from the shopping trolley at the till then repacked in the customer's bags.*

HOW TO READ BARCODES

A universal barcode is made up of a sequence of vertical lines uniformly 0.5 mm wide. The black bars represent the digit 1 in the binary system, white bars represent zero. Depending on the precise arrangement of the bars, they comprise codes of 8 (EAN-8) or 13 (EAN-13) digits which, reading from left to right, encode information on the country or region of origin ('50' denotes the UK), the product, the manufacturer and a control number confirming that the barcode is valid.

More recent refinements to the bar-code system include 'stacked' and 'matrix' barcodes that can hold additional product information. Meanwhile, scanners have become cordless, allowing them to read barcodes from up to 10 metres away, while some are even fitted with cameras.

Digital gateway *The opening screen of artist Scott Blake's website is a visitor number counter in the form of a giant bar-code.*

Credit cards 1950

The credit card was a product of absent-mindedness. In 1949 top business executive Frank McNamara, head of the Hamilton Credit Corporation, was dining in the exclusive Major's Cabin Grill restaurant near the Empire State Building in New York. At the end of the meal, as he came to settle up, McNamara discovered that he had left his wallet at home. Fortunately, his wife spared his embarrassment by paying the bill, but the incident set him thinking about creating a card that could be used to dine out without having to carry cash. On 8 February, 1950, McNamara and his business associate Ralph Schneider founded the Diners' Club, the world's first credit-card company. Their very first transaction was, fittingly, at Major's – an event jokingly known in the credit-card industry as the 'First Supper'.

The cardboard membership card issued initially provided a very limited service: subscribers could use it in 27 New York restaurants and had to settle their bill in full within 30 days. But before long, more and more restaurants started to accept the card. By 1951 there were already 20,000 card-holding members of the Diners' Club and 400 participating eateries. As it spread to other US cities and then abroad (from 1955), hotels, car-hire companies and florists also signed up.

The rise and rise of credit

Even before this surge in popularity, other financial institutions had spotted the potential. American Express launched its first bank card in 1958; within five years, more than a million of them were being used in some 85,000 establishments worldwide. The same year, the Bank of America brought out the Bank-Americard, renamed Visa in 1976. The range of services increased to include deferred payment, with interest charged on the loan, then the first ATMs appeared making it easier to draw out cash – for a fee.

By 2000, credit card purchases added up to some £130 million worldwide. There are now thought to be around 10,000 credit-card transactions every second. The advent of magnetic swipe cards and then 'chip-and-pin'-activated debit cards have cut overheads in the retail banking sector and raised profits by a factor of 25 per cent. The original credit card, Diners' Club, disappeared in 2009.

Shopping spree
Credit cards have made shopping easier than ever and this smiling young woman and her credit card (below) epitomise today's easy-going attitude towards credit. But after the financial crisis of 2008 many have blamed easy credit for fuelling a 'boom-and-bust' economic cycle by encouraging people to live beyond their means.

First of many
The original Diners' Club credit card issued (above). In 1957 a couple from Lancaster, Pennsylvania, made history by taking a round-the-world tour equipped only with a Diners' Club card and their airline tickets.

THE FIRST WIVES' CLUB

In 1959 Diners Club introduced a Women's Division, giving the wives of card-members the opportunity for cash-less shopping in New York boutiques, beauty salons and health spas.

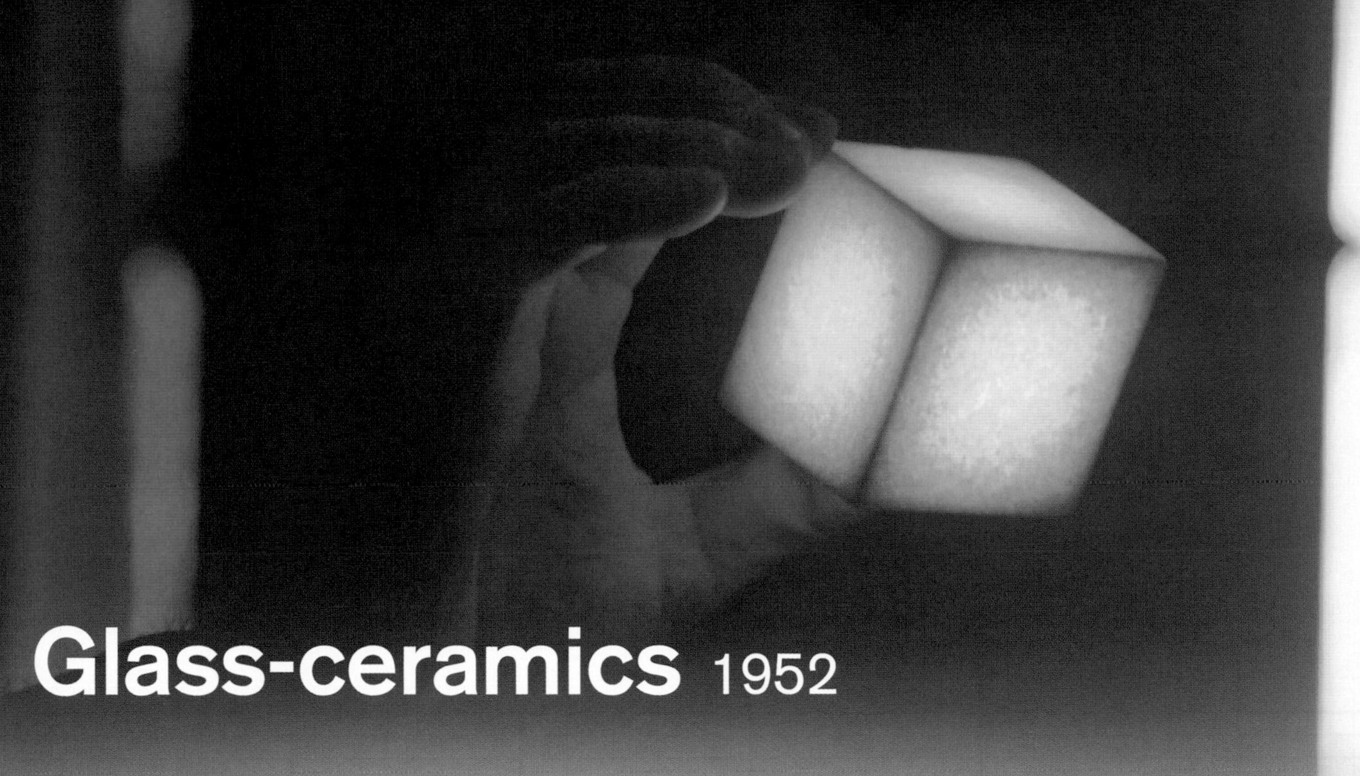

Glass-ceramics 1952

As Stanley Donald Stookey opened the kiln to take out the piece of glass he had put in several hours earlier, his first surprise was that it had become opaque. But a bigger surprise awaited him: as he lifted the piece out, it slipped from the tongs, fell to the floor ... and bounced. The year was 1952 and the place the Corning Glass Works in New York State.

Stookey soon realised that the kiln's thermostat had malfunctioned and that the temperature had risen far higher than he intended – to 900°C rather than 600°C. The new shock-resistant material he had just accidentally created came to be called glass-ceramics. He patented the invention in 1956.

Unique properties

Technically, glass is an 'amorphous solid', which means there is no long-range order in the position of its atoms: it is a liquid congealed so rapidly by supercooling that the atoms have no time in which to form a regular structure. By overheating the glass, Stookey had stimulated the controlled formation of microscopic crystals – that is, structures with a regular arrangement of atoms. Unlike ordinary glass, glass-ceramics can withstand huge fluctuations in temperature and sudden shocks. At the same time, it retains the durability and permeability of glass to infrared radiation.

From the kitchen to outer space

One of the first applications of glass-ceramics, by the US military, was in nose cones for the radomes enclosing supersonic guided anti-aircraft missiles. Its heat-resistance and transparency to radar waves made it ideal for this purpose. In 1958 Corning introduced CorningWare, a range of oven-to-table cookware made from Stookey's 'Pyroceram'. Glass ceramics were also used for heating elements on electric cookers. In 1975 researchers created the first transparent glass-ceramics, now widely used in lasers, fibre-optic cables and mirrors for space telescopes.

ENTOMBED IN GLASS

Long-term storage of highly radioactive nuclear waste requires it to be stabilised in a form that will not deteriorate and release contamination. This is achieved by vitrification. After being heated and reduced to a fine powder, the waste is bonded into a glass-ceramic matrix that will not break down for hundreds of thousands of years. While it is still fluid, the vitrified waste is poured into cylindrical stainless steel containers and slowly cooled before being sealed in a second metal drum. These are then stockpiled in hardened underground concrete bunkers.

Magic cube
The glass-ceramic coating on this cube (top) prevents the cube shattering or becoming misshapen when heated to 1,200°C. The edges of the cube remain impervious to heat, so it can be picked up by hand straight from the kiln.

Efficient conductor
More than 90 per cent of the heat generated by the element beneath this ceramic hob is transmitted through the glass to the base of the pan.

THE PACEMAKER – 1951

Ensuring a healthy heartbeat

While studying the effects of hypothermia, Canadian researchers hit upon the idea of stimulating the heart with electrical pulses. This was the first practical step towards introducing the cardiac pacemaker. Nowadays, implanting a pacemaker has become a routine medical procedure, which has saved thousands of lives.

Dr Wilfred G Bigelow had been studying hypothermia at the Banting Institute in Toronto since 1946. One day in 1949, his team cooled the body temperature of a dog to around 22°C, when the animal's heart stopped beating. Cardiac massage failed to restart it. In desperation, Bigelow poked it with an electric probe, which produced a strong contraction. He poked again, with the same result. After some minutes of stimulation, the team observed that the dog's blood pressure had returned to normal.

Bigelow realised straightaway the potential of applying an outside, artificial stimulant to induce the heart to beat. Could a repeated electrical pulse be used to compensate for irregularities in the sinoatrial node, the impulse-generating tissue located in the right atrium of the heart? He contacted the National Research Council of Canada and secured the services of John Hopps, an electrical engineer. Together with Hopps and Dr John Callaghan, a colleague from the hypothermia programme, Bigelow set to work to create a practical heart pacemaker.

Miniaturisation pioneer
In the 1960s medical technician Geoffrey Davies of St George's Hospital in London was at the forefront of research to devise miniaturised, low-voltage pacemakers, like this one (below).

Controlling the heart rate

As early as 1932 Dr Albert Hyman, a New York physiologist, had constructed a hand-cranked device emitting intermittent electrical stimuli, which he used to restart the stopped hearts of small mammals such as rabbits and guinea pigs. But apart from this prototype artificial pacemaker, no other research was conducted in this area until the Canadian team began their work. Their aim was not just to replicate Hyman's success in getting a stopped heart to start beating regularly again, but also to use an external stimulus to control the heart rate of patients suffering from conditions such as bradycardia (an abnormally low cardiac rate).

For their early experiments on dogs, Bigelow and his team attached a catheter electrode directly to the myocardium (the cardiac muscle). But they soon found that they could stimulate the sinoatrial node by means of a probe equipped with two electrodes introduced into the heart through the jugular vein.

In 1951 they unveiled the world's first external electronic pacemaker. Housed in a metal box almost 30cm long, it comprised an electrical circuit that ran off the regular domestic mains electricity supply at 60Hz. The device was immediately taken up by several hospitals in the United States, and in

Leading a full life
A 1961 issue of the American photo-journalism magazine Life *devoted a report to this patient (left) who had been fitted with a new Medtronic external pacemaker. She was photographed actively bowling and swimming.*

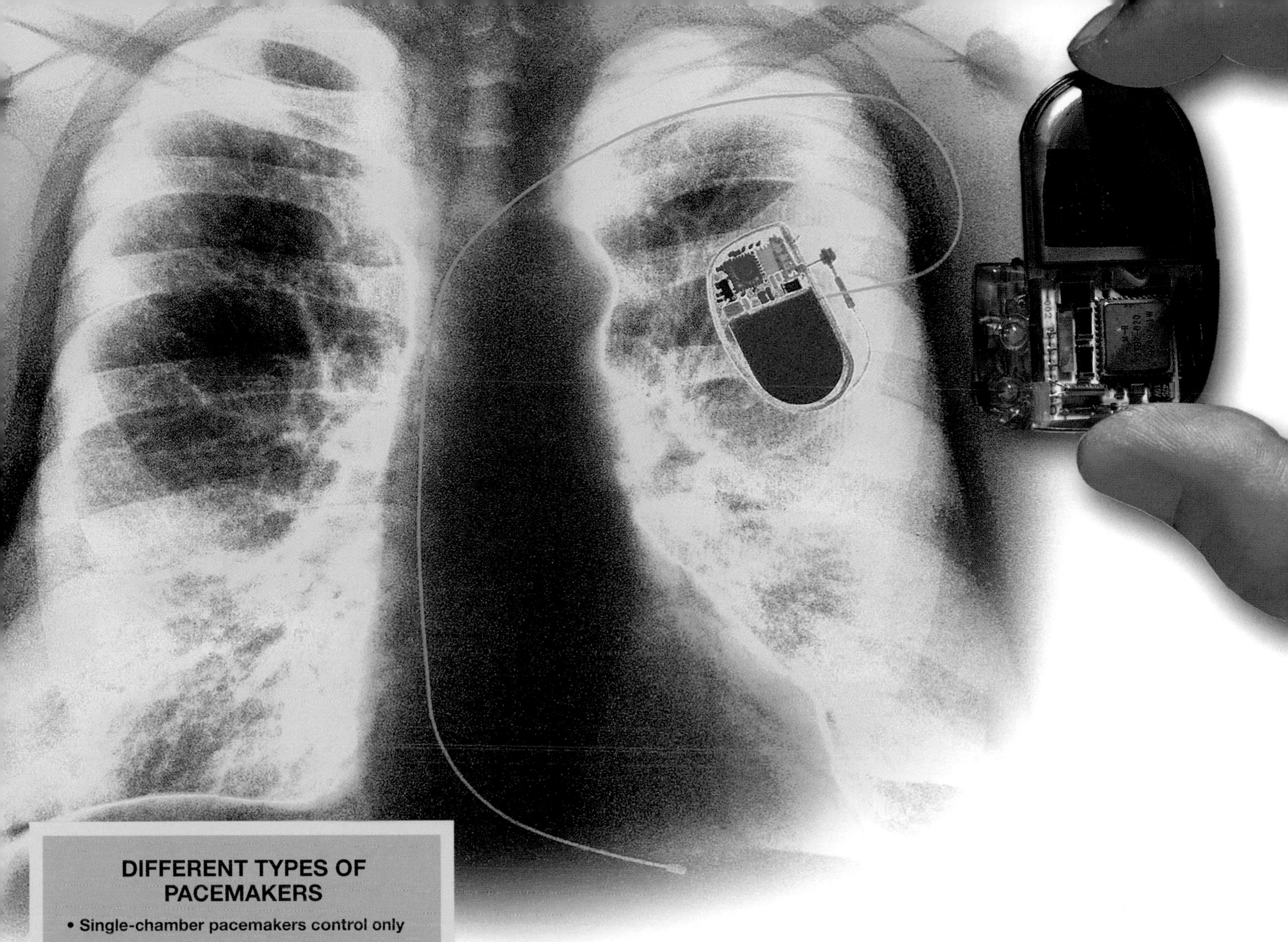

> **DIFFERENT TYPES OF PACEMAKERS**
>
> - Single-chamber pacemakers control only a single chamber of the heart, whereas dual-chamber pacemakers control both the ventricles and the atria.
> - Rate-responsive pacemakers automatically adjust the pacing rate to the patient's level of physical activity.
> - Multisite pacemakers are used for patients suffering from dilated cardiomyopathy (DCM), a condition that impairs the ability of the left or right ventricles to contract efficiently. These pacemakers help to resynchronise the ventricular contractions.

1952 the renowned American cardiologist Paul Zoll used it successfully in the treatment of several of his patients.

The first heart implants

Six years later, in Stockholm, Dr Ake Senning conducted the first implantation of a cardiac pacemaker. The model in question was battery-powered and had been developed by Rune Elmquist, an engineer from the Siemens-Elema company, who drew on the miniaturisation of electronic circuits pioneered by the aeronautical industry to produce a device around the size of an ice-hockey puck. Senning implanted it in the thorax of a 43-year-old man with bradycardia. The operation was far from trouble-free – the pacemaker proved faulty and had to be replaced six hours after the original surgery. But against all the odds the patient survived to live into his eighties, having gone through 30 pacemakers by the time he died.

Pacemaker battery technology improved greatly in the 1960s, particularly as a result of the work of the American biomedical engineer Wilson Greatbatch. Yet it was the 1970s before pacemakers with a lifetime of 5 to 10 years or even more became available, thanks to the invention of lithium batteries, which are much longer-lasting than mercury ones. Modern pacemakers are externally programmable, allowing the cardiologist to select the optimum pacing modes for individual patients and to monitor patients remotely. The body of these devices is now very small, measuring only around 4 centimetres on average and weighing just a few grams.

Miniature lifesaver
A coloured X-ray of a patient's chest shows a fitted cardiac pacemaker clearly visible above the ribcage; the green lead connects it to the heart. The device is electronic and battery-powered – an unfitted example is shown at top right.

ULTRASOUND SCANS – 1952

Looking inside the human body

The underwater tracking system called sonar works by sending out sound waves and analysing the reflected signals. The medical world adopted this technology as a way of investigating inside the body without invasive surgery. Ultrasound scanning proved a great success, especially in the field of obstetrics.

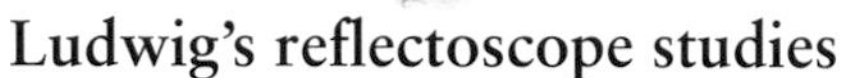

After the Second World War, the Allies declassified sonar. As a defence system during the war sonar had been top-secret: it played a vital role in the Battle of the Atlantic, detecting German submarines by emitting pulses of sound and listening for echoes. Non-military application of the technology had begun as early as 1942, when the Austrian neurologist Karl Dussik published a paper entitled 'On the possibility of using ultrasound waves as a diagnostic aid'. Dussik's work was the first to raise the possibility of differentiating different body tissues by subjecting them to ultrasound transmission.

LIMITS OF ULTRASOUND

Ultrasound scanning cannot be used to study the skeleton, since the waves are absorbed by, rather than bouncing off, bone and gases, including air.

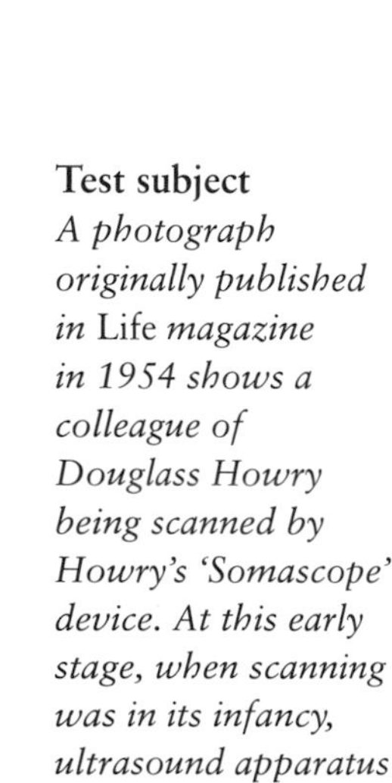

Test subject
A photograph originally published in Life *magazine in 1954 shows a colleague of Douglass Howry being scanned by Howry's 'Somascope' device. At this early stage, when scanning was in its infancy, ultrasound apparatus was remarkably unwieldy. The subject's kidney can be seen on the Somascope screen (top right in the picture).*

Ludwig's reflectoscope studies

A few years later, at the US Naval Medical Research Institute in Bethesda, Maryland, George Döring Ludwig began experiments on animals using a reflectoscope – an instrument based on the same principles as sonar, which was used in industry to test for flaws in materials. Ludwig's tests involved detecting the presence and position of foreign bodies, especially gallstones, in animal tissues and organs, transcribing onto a curve the variations in amplitude of the reflected signal according to the depth beneath the skin of the structures he encountered. He published his findings in a detailed report in June 1949, where he suggested that ultrasound was useful not just for detecting kidney stones or gallstones but that it could also ultimately be used to locate tumours.

The first images

Ludwig's paper effectively paved the way for ultrasound scanning. Three years later, in 1952, came the publication of seminal studies conducted by John Wild and John Reid of the University of Minnesota and by Douglass Howry, a radiologist from Denver. Reid and Wild employed ultrasound to measure the variations in density of human tissues and then applied this data to the study of breast tumours. With the help of an electrical engineer, they built a scanner that could visualise the results in two dimensions. Two levels of luminosity showed the differential sonic energy reflection from healthy and malignant tissue. Meanwhile, Douglass Howry constructed his own pulse-echo ultrasonic scanner to obtain accurate cross-sectional images of soft-tissue structures.

Ultrasound technology progressed rapidly thereafter. In 1954 the Scottish gynaecologist

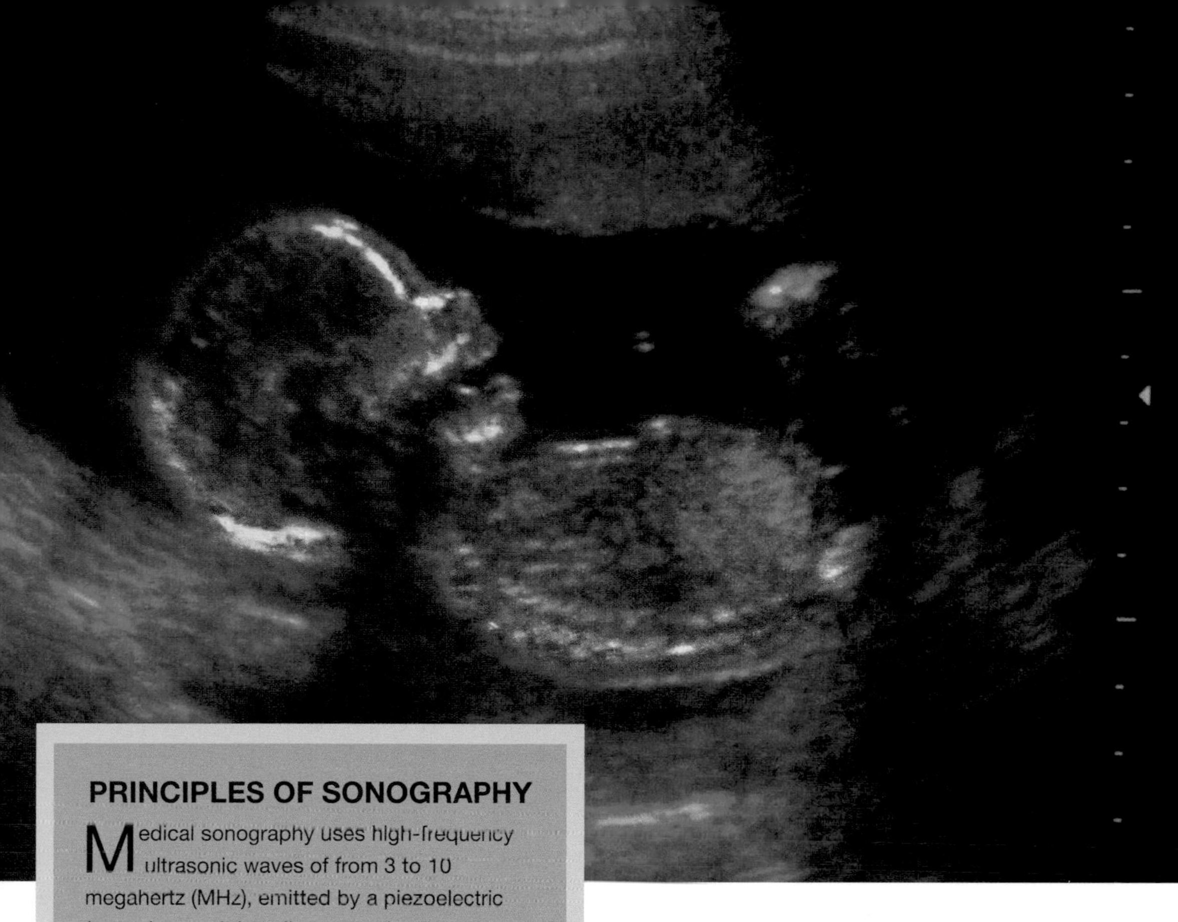

Foetal scan
Anomaly (or foetal morphology) scans take place between the 18th and 20th weeks of pregnancy, by which time the foetus is well developed, to check that the baby is developing normally; at this stage the scan can also discern the baby's sex. In this scan (left), the profile is well defined, clearly showing the open mouth, one leg and the spinal column.

PRINCIPLES OF SONOGRAPHY

Medical sonography uses high-frequency ultrasonic waves of from 3 to 10 megahertz (MHz), emitted by a piezoelectric (quartz) crystal that vibrates under the influence of electrical pulses. The catheter both transmits the ultrasonic waves and receives the resultant echoes. Waves are produced in liquids – the gel that is applied to the skin prevents the signal from becoming dissipated – and are partially echoed whenever they encounter an interface between tissues. As the echo pulses return to the body's surface they are picked up by a transducer and converted back into electrical pulses that are then processed by the system and formed into an image.

and obstetrician Ian Donald was appointed as Professor of Midwifery at the University of Glasgow. Formerly an officer with the RAF, Donald was already familiar with radar when an encounter with Wild led him, with help from his engineer assistant Tom Brown, to develop ultrasound scanning for his own field. In 1958 Donald published a groundbreaking study that included pictures of ovarian cysts, fibromas and the uteruses of pregnant women. The next year, he produced a detailed image of the head of a foetus in the womb.

Studying foetuses in real time

In the 1960s ultrasound scans were taken using a hand-held probe. Over the following decade the development of automated scanners, which enabled the foetus to be examined in real time, opened up new horizons in the monitoring of pregnancy. Greyscale imaging greatly improved the level of detail on scans. Also, the advent of Doppler ultrasound scanning, which registered the changes in frequency of sound waves transmitted by an object in motion (the Doppler Effect), enabled blood to be evaluated as it flowed through blood vessels. Since its inception, the technology has undergone major advances – colour and 3-D images are now the norm – and ultrasound tests have become routine procedure, especially in cardiology and obstetrics. Almost all pregnant women in the developed world undergo a series of antenatal scans.

3-D image
3-D sonography was developed at the end of the 1980s to improve detection of foetal anomalies. Its accuracy is evident in this comparison of a 3-D scan image (below left) with a photo in profile of the newborn infant.

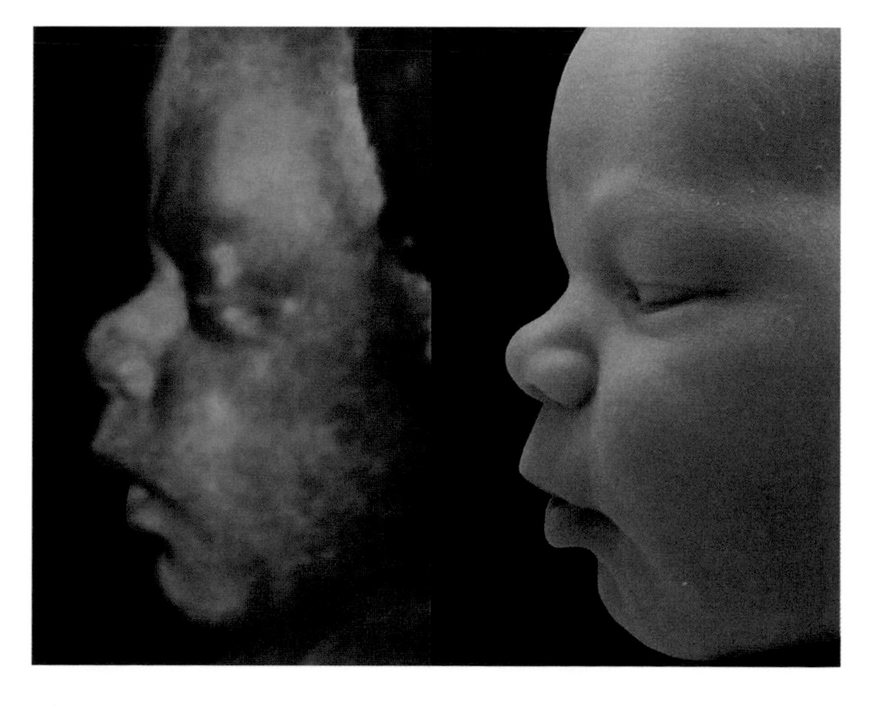

The H-bomb 1952

Fearsome weapon
Russia's 'Tsar Bomb' (right) was put on show for the first time in Moscow in 1992, after the end of the Soviet era. It was 8m long, 2m wide and weighed around 27 tonnes.

On 1 November, 1952, Eniwetok Atoll in the Marshall Islands was the scene of the first explosion of a thermonuclear or hydrogen bomb. The energy released by this bomb, nicknamed 'Mike', was equivalent to 10.4 megatons of TNT – 500 times the yield of the plutonium bomb detonated over Nagasaki in August 1945. The H-bomb test in the Pacific ushered in an age where the destructive power of nuclear weapons seemed limitless.

Inspiration from the stars

American physicist Edward Teller spent a decade developing the hydrogen bomb. The atomic bomb had worked on the principle of fission – that is, the nucleus of the atom is split by being bombarded by neutrons, causing a chain reaction. The new thermonuclear bomb relied on the nucleosynthesis of light atoms – namely, hydrogen – to re-create the reaction that takes place in and powers the stars. Fusion of the hydrogen nuclei, which releases a phenomenal amount of energy, was triggered by the explosion of a fission device raising the temperature within the bomb to several million degrees.

Although most other American physicists had fought shy on moral grounds of collaborating with Teller on the H-bomb, they were galvanised into support in 1949 when the Soviet Union exploded its first atomic bomb. Then, as the Cold War progressed, each side tried to increase the yield of its nuclear weapons, while at the same time reducing the physical size of warheads in order to mount them on Intercontinental Ballistic Missiles (ICBMs). Modern nuclear weapons typically have a yield of between 100 kilotons and 1.5 megatons. Still only a handful of countries have the technological capability to manufacture H-bombs: the USA, Russia, Britain, France, China, India, Pakistan and Israel.

Dawn of destruction
The detonation of the first American H-bomb at 0715 on 1 November, 1952, produced a huge fireball that soon transformed itself into a towering column of smoke and debris, surmounted by a vast white mushroom cloud.

BIG BANG

The Russian AN602 hydrogen bomb, nicknamed *Tsar Bomba*, was the most powerful nuclear weapon ever exploded. Tested on Novaya Zemlya Island in 1961, it had a yield of over 50 megatons.

A USELESS DETERRENT

The first American H-bomb was too heavy to have been deployed against the Soviet Union. It used deuterium ('heavy hydrogen') as its fissile material, which needed to be cooled to –250°C to keep it in a liquid state and prevent it from vapourising. The refrigeration system required to achieve this meant that 'Mike' weighed a massive 65 tonnes and no bomber could carry it. The USSR tested its first hydrogen bomb on 12 August, 1953. Yielding around 400 kilotons, the Soviet weapon was almost certainly less powerful than the American bomb, but had the advantage of being small enough to fit in a strategic bomber's bomb-bay.

The Prioré Machine 1953

Immediately after the Second World War, Antoine Prioré, an Italian-born electrical engineer who fled his home country and settled in Bordeaux, began a series of experiments aimed at curing cancers by subjecting them to a strong magnetic field generated by a large machine of his devising. He claimed to have obtained positive results from tests on rats in 1953. Eleven years later, a child with cancer who was treated by the machine was given a clean bill of health.

Anti-cancer machine *Although he refused to release the records of his experiments with his magnetic device (left), Antoine Prioré (inset) claimed that its efficacy depended upon a combination of the body mass of the patient and the length of time that the patient was exposed to the magnetic field during treatment.*

Although Prioré gained the backing of some renowned scientists, of France's political establishment and even of the World Health Organisation, his claims for his device are still the subject of great controversy, as he refused to disclose how it worked. Since his death in 1983, laboratory studies have shown that certain magnetic waves can have an effect on animal cells, but it remains uncertain whether they are ultimately beneficial or harmful. Most medical researchers today regard electromagnetic therapy as a pseudoscience.

A HIGHLY CHARGED CONTROVERSY

The 1990s saw a revival of interest among doctors in the effect of magnetic fields – specifically those generated by high-tension overhead power lines and microwave transmitting masts. At the time of writing no conclusive proof has emerged to indicate whether long-term exposure to these waves is either harmful or completely safe. In consequence, experts have adopted the precautionary principle and advise people to limit their exposure.

Pressure cookers 1953

The first attempt at making a pressure cooker was French physicist Denis Papin's 'steam digester' of 1679, which he used to extract gelatin from animal bones. An American firm, National Presto Industries, revived the idea in 1939, exhibiting it at the New York World's Fair that same year, but the war and safety concerns delayed its introduction. The world's first mass-produced domestic pressure cooker was the French 'Cocotte-Minute', launched by SEB (Société d'Emboutissage de Bourgogne) in 1953. Nowadays, many people's kitchens contain one or another model of this cooking appliance. The food is placed in an inner metal basket, which sits inside the airtight outer shell, and is then covered with water. During cooking the water evaporates, increasing the pressure within the vessel and hence the temperature, with the result that cooking time is reduced to around a third or even a quarter of what it would be using a conventional pan. A safety valve releases the steam if too great a pressure builds up, while an automatic locking device prevents the cover from being removed before the vessel is completely decompressed.

Faster food *Most modern pressure cookers work at 15psi above atmospheric pressure, where water boils at 122°C. This makes them very effective sterilisers, for instance for jam-making or baby bottles.*

THE STRUCTURE OF DNA – 1953

Revealing the double helix

James Watson, Francis Crick and Rosalind Franklin's discovery of the structure of DNA opened up new horizons in the life sciences. As researchers gained a better understanding of the mechanisms of heredity and the functioning of the cell, the discipline of molecular biology was born.

Molecular model
Francis Crick (on the right) and James Watson in the Cavendish Laboratory in Cambridge in 1953, the year of their great breakthrough. Watson later recalled: 'When we saw the answer, we had to pinch ourselves. We realised it probably was true because it was so pretty.'

Scientists had known since 1924 that chromosomes contained genes and that these, in turn, contained two acids. One of these was deoxyribose nucleic acid (DNA for short), which was vital for cell growth and in transmitting inherited traits. The other was ribonucleic acid (or RNA), present in the cell nucleus and in the cytoplasm surrounding it; the precise role of this compound was still unclear.

As early as 1944, the German Max Delbrück and the Austrian Erwin Schrödinger surmised that the secret of life was to be found in the structure of chromosomes and genes, and specifically in DNA. In a popular science book he published that year, entitled *What is Life?*, Schrödinger described in theory how the storage of genetic information might work. Yet Delbrück and Schrödinger were physicists, and it still remained for biologists to explain exactly what DNA was composed of.

FROM NUCLEIN TO DNA

In 1869, in the course of research into the composition of cell nuclei, the Swiss biochemist Johannes Friedrich Miescher extracted from the nuclei of white blood cells a substance he called 'nuclein' (which we now refer to as nucleic acid) made up of proteins and a phosphate-rich compound. He also showed that nuclein was present in the spermatozoa of several species and surmised that this substance might well play a role in heredity. In 1889 Richard Altman removed the protein from nuclein in yeast cells and obtained an organic phosphoric acid that he named nucleic acid. In the 1920s, research by Albrecht Kossel – who had won the Nobel prize for physiology or medicine in 1910 for his work on cell chemistry – and by the Russian-American biochemist Phoebus Levene gradually revealed precisely what this acid was composed of: phosphorus, sugar (deoxyribose) and four nitrogenous bases – adenine, guanine, cytosine and thymine. As a result, the compound was given the scientific name of deoxyribose nucleic acid, now usually abbreviated to DNA.

X-ray diffraction

In 1947 four researchers working in Britain – the Englishman William H Bragg, the German Max von Laue, New Zealander Maurice Wilkins (a former physicist turned biologist) and the Austrian Max Perutz – introduced a new method for investigating the structure of living organisms: X-ray diffraction. Diffracted by a crystal and recorded on photographic film, the X-rays produced a different pattern according to which elements – carbon, sulphur, hydrogen and so on – were present in proteins. The technique enabled the researchers to identify the constituent elements of each protein, a crucial step in understanding DNA. It was soon augmented by another vital new tool, paper chromatography developed by the Austrian Erwin Chargaff, working at Columbia University in the USA.

THE LOST PIONEER

While Watson and Crick became household names for their work on DNA, Rosalind Franklin remains almost unknown. Yet it was she who produced the remarkable X-ray diffraction images that enabled the Cambridge team to determine the structure of the molecule. Trained as a physical chemist, Franklin had a rocky professional relationship with Maurice Wilkins, the head of her research unit at King's College in London. She resented his exploitation of her experimental expertise and his treatment of her as a lab assistant rather than his scientific peer. Watson also painted a distorted picture of her role as purely subsidiary in his account of the discovery of DNA, *The Double Helix* (1968). Franklin died of ovarian cancer in 1958 at the age of 38. As the Nobel prize is not awarded posthumously, she was not included in the prize that acknowledged Watson, Crick and Wilkins in 1962.

Vital images *X-ray diffraction was an advanced technique to determine the structure of crystalline compounds. Rosalind Franklin (top) was an acknowledged expert in the field. Above: one of her images of DNA.*

Spiral staircase

These techniques gave a detailed new insight into the mysteries of DNA. Two young Cambridge researchers, Francis Crick and James Watson, began working on the problem, in parallel with a team at King's College, London, led by Maurice Wilkins and his chemist colleague Rosalind Franklin, a renowned crystallography expert. The 35-year-old Crick, himself an ex-physicist, was writing a thesis on haemoglobin at the time, while Watson, 23, had obtained his doctorate in genetics the year before. They had both studied under the eminent American chemist Linus Pauling, who had been grappling with the structure of DNA for some years. When the pair met at the Cavendish Laboratory in Cambridge, they soon discovered a mutual obsession with the problem.

Yet Crick and Watson were far from being pioneers in the field. From 1944 onwards, Scottish biochemist Alexander Todd had ascertained that the chain of molecules (nucleotides) that comprise DNA were consistently formed by sugar molecules and phosphate groups that attached themselves to four known chemical bases of DNA – adenine, guanine, cytosine and thymine (commonly denoted by the letters A, G, C and T). Chargaff had also discovered this same structure.

Crick and Watson noticed that these bases did not join together haphazardly: adenine

Jigsaw pieces *Left: Coloured aluminium models made by Crick and Watson to represent the chemical bases of DNA.*

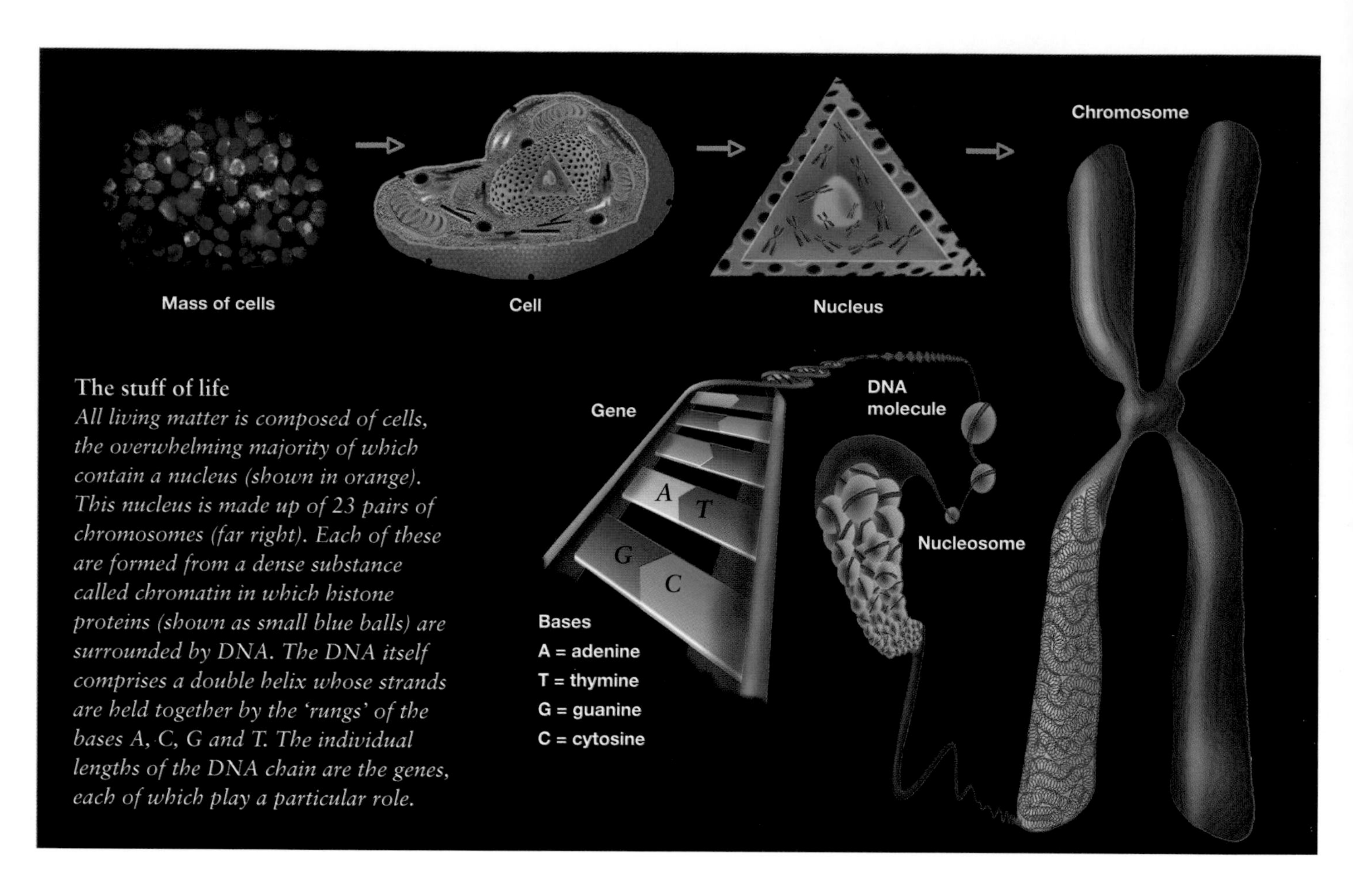

The stuff of life
All living matter is composed of cells, the overwhelming majority of which contain a nucleus (shown in orange). This nucleus is made up of 23 pairs of chromosomes (far right). Each of these are formed from a dense substance called chromatin in which histone proteins (shown as small blue balls) are surrounded by DNA. The DNA itself comprises a double helix whose strands are held together by the 'rungs' of the bases A, C, G and T. The individual lengths of the DNA chain are the genes, each of which play a particular role.

The double helix
A graphic representation of the structure of DNA shows the four different coloured 'rungs' that join the two 'upright' chains of phosphates. The order of the rungs follows a pattern: adenine (A) pairs with thymine (T), while guanine (G) pairs with cytosine (C), in an infinite number of different combinations. This variety means that the genetic code of every individual is unique.

always paired with thymine and guanine with cytosine. Yet simply knowing what DNA was made of was not the same as understanding its structure. An X-ray diffraction image taken by Rosalind Franklin, known as 'Photo 51', proved a key catalyst to their discovery of how the edifice fitted together. But cracking it entailed long hours with home-made cardboard models trying to extrapolate from the few fragments they were sure of in order to build up a coherent picture of the whole. As Watson later wrote: 'I noticed that an A-T pair, linked by hydrogen bridges, had the same shape as a G-C pair.'

The final breakthrough came when they realised how the chemical bonds between the base pairs worked. The resulting model of the DNA molecule they proposed resembled a ladder, whose two 'uprights' were formed by polynucleotide chains of phosphates and sugars. The rungs holding the chains together were the base pairs. The uprights of the ladder were not straight, but rather were twisted like ropes into a shape resembling a spiral staircase, called a 'double helix'.

The driving force behind replication

Scientists still had much work to do. The elegant model of the structure of DNA did not explain how it ensured the survival and replication of cells, nor the role it played in heredity – that is, in the transmission of genes.

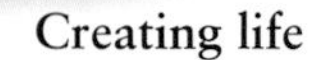

Creating life
As DNA molecules 'unzip', the four bases that form the links between strands (here shown as different coloured balls) are exposed. The unzipped strands then link up with spare nucleotides to form new strands.

THE VERSATILE PROTEINS

Proteins perform a wide variety of different functions within living organisms. For example, enzymes speed up chemical reactions, haemoglobin carries oxygen within the blood, keratin makes up hair and nails, while antibodies help to fight infection. The many separate proteins that perform these roles are all composed of chains of small molecules known as amino acids.

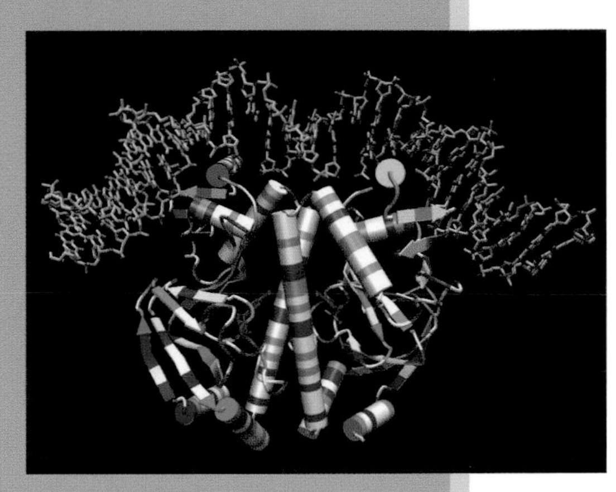

Cause and effect
Effectors are molecules that bind to proteins and activate certain genes according to the cell's needs. The illustration above shows an effector (striped cylinders) attaching itself to a DNA molecule (the green and orange lace structure), which it bends to specific sites.

Nor could it account for the role of the other nucleic acid, RNA. Further research by Crick and Watson, this time conducted separately, provided answers to these questions by 1961. Crick discovered that each group of three bases (or 'triplets') on a single DNA designates the position of a specific amino acid on the polypeptide chain of a protein molecule. He also helped to determine the base triplets that code for each of the 20 amino acids normally found in proteins, which enable the cell to replicate. Watson highlighted the role played in protein synthesis by RNA, in which thymine is replaced with the base uracil, and which carries coding instructions from DNA to structures called ribosomes, which are the components of cells that make proteins out of amino acids. But how could just four bases – A, G, C and T – code for 20 different amino acids? The answer was that they combine in different sequences, such as AGTCC, AGGA, AGCT, AGTGC, ACTG and so on, and each of these combinations produces a different amino acid.

All this helped to reveal the secret of replication, a process that begins at a given stage in the life of a cell with the unzipping of the two helicoidal strands that make up DNA. These link up with a series of new nucleotides transmitted by the messenger RNA to form new strands. This insight represented a major

WHAT ARE GENES MADE OF?

In the early 1930s, at the Rockefeller Institute in New York, Oswald Avery began research into an intriguing phenomenon first identified by the British biologist Frederick Griffith in 1928 in the course of work on pneumococci (the bacteria that cause pneumonia). The phenomenon was called bacterial transformation, the ability of a bacterium to change from one strain into a different strain, thereby altering its function. In 1944 Avery and his assistants Colin MacLeod and Maclyn McCarty succeeded in demonstrating that DNA was the factor that brought about this transformation, also known as 'genetic recombination'. Avery's work excited the interest of some later researchers, notably James Watson, but was ignored by most biologists, who remained convinced that only proteins could carry genetic information. An experiment conducted in 1952 by Alfred Hershey and Martha Chase settled the matter once and for all: by labelling proteins and the DNA of a virus with different radioactive isotopes, the researchers proved conclusively that it was DNA that contained the genetic information necessary for the formation of new strains of the virus. This important discovery helped to pave the way for Watson and Crick's breakthrough the following year.

step forward in molecular biology, and its authors were awarded the Nobel prize in 1962. Yet this was far from being the end of the story.

Opening the door to knowledge

The discovery of DNA was a decisive turning-point in biology, as it opened the way for a better understanding of living organisms, which henceforth could be studied and described at the molecular level. Crick, Watson and Wilkins' breakthrough prompted a whole raft of new questions, as scientists began to ask at what stage the cell replicated, what mechanism regulated this process, and whether this mechanism operated independently or was influenced by environmental factors?

The decade following Watson and Crick's discovery was marked by a series of other key discoveries, including the deciphering of the genetic code by Nirenberg and Khorana. The explanation of the stages by which proteins are synthesised and the way in which genes control this process won François Jacob and Jacques Monod a Nobel prize in 1965. As a result, gene technology boomed in the 1970s.

From the limited context within which the structure of DNA was revealed in 1953, solely with regard to the question of heredity, the field of molecular biology has expanded to encompass every aspect of living organisms, including genetic modification. The mapping of the genomes of humans, other animals and plants has enabled biologists to produce food crops capable of resisting drought or parasites and to use gene therapy to treat hereditary diseases. Other common applications include DNA profiling for paternity tests and genetic fingerprinting in crime detection. Many areas of study – most notably genetically modified foods, cloning and stem-cell research – have raised ethical and environmental concerns, but what remains beyond dispute is the increasing role that genetic science will play in years to come.

THE GENETIC CODE

DNA is often likened to an instruction manual, since it contains all the information necessary for creating the proteins that go to make up an organism and ensure its proper functioning. But how exactly is this information encoded within DNA? There is a direct correspondence between the sequence in which the bases of DNA link up and the way in which the different amino acids that constitute proteins are organised. This system of correspondence is known as the genetic code. Its discovery by the biochemist Marshall Nirenberg and Har Gobind Khorana earned them the Nobel prize for physiology or medicine in 1965. Proof of the universality of the genetic code – the correspondence between the DNA bases and amino acids is the same in all living organisms – has had major implications for the development of biotechnology, since it is this that makes it possible to transfer genetic material from one organism to another.

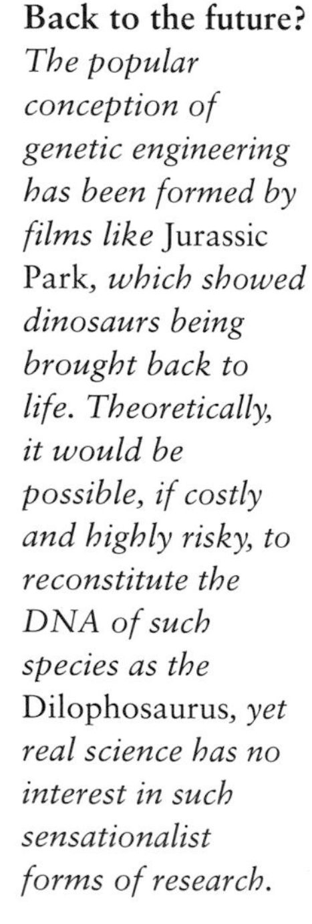

Back to the future? *The popular conception of genetic engineering has been formed by films like* Jurassic Park, *which showed dinosaurs being brought back to life. Theoretically, it would be possible, if costly and highly risky, to reconstitute the DNA of such species as the* Dilophosaurus, *yet real science has no interest in such sensationalist forms of research.*

Close cousin

A female bonobo (a species of chimpanzee) demonstrating her ability to recognise signs in ongoing research into language acquisition at the University of Georgia in Atlanta. Humans share 99 per cent of their genes with other great apes.

MEDICINE AND GENETICS

The huge advances made in molecular biology since the 1950s have enabled great strides forward in medicine. In the case of hereditary diseases, for instance, doctors now understand how the alteration of a gene by modifying the way the bases link up can prevent the synthesis of a particular protein, or stimulate the production of a defective one, which may have serious repercussions for the health of an individual. Deeper understanding of the mechanisms at work in these diseases has prompted the development of more effective remedies.

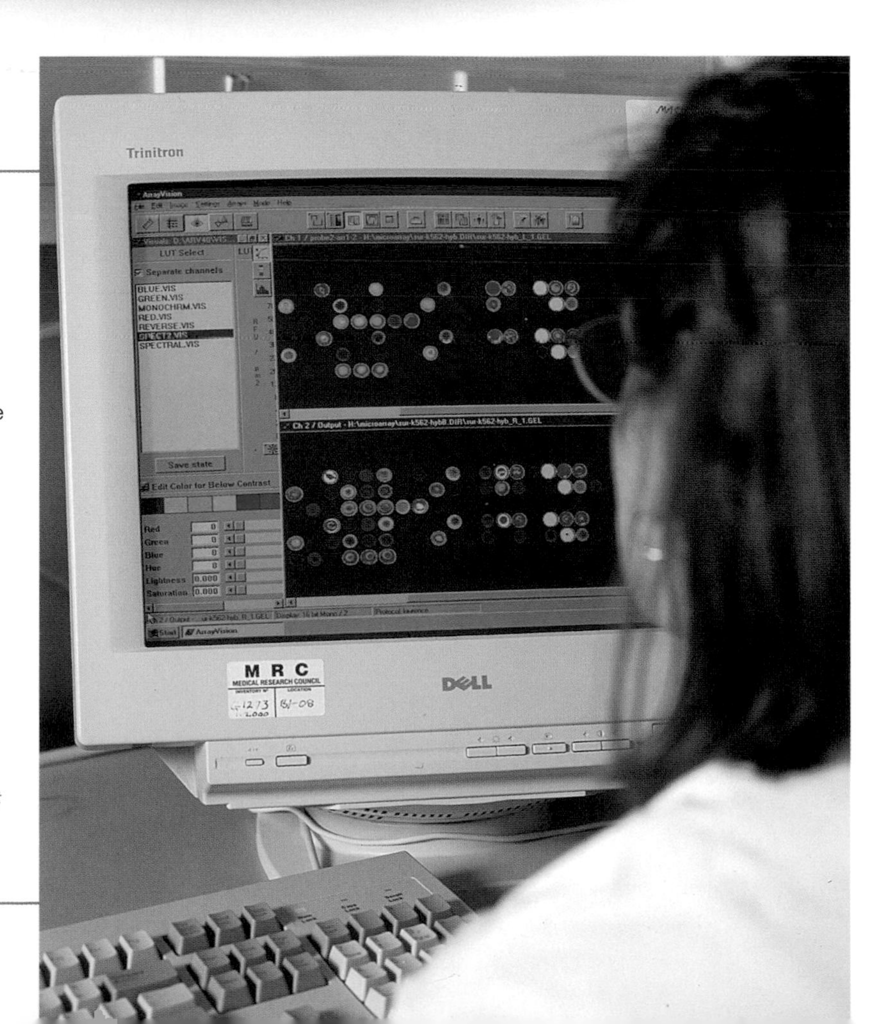

Bringing benefits

A research scientist in Cambridge at work on the Human Genome Project (right). This major study, which began in 1989, was completed in 2003 with the release of a complete map of the 20,000-plus genes that make up the human genome. The project has gone on to provide essential data for medical research and biotechnology.

POLAR EXPLORATION

Charting the poles

The beginning of international cooperation on polar research can be dated to 1957, which was named both International Geophysical Year and the Third International Polar Year. Formerly the preserve of a select band of intrepid adventurers, polar exploration was about to change radically.

Controversial claim
Robert Peary's Arctic team in 1909 included 17 Inuit hunters and sled-drivers. Most experts now think that Peary never reached the North Pole.

Trapped in ice
The hull of the Fram, *Fridtjof Nansen's Arctic exploration vessel, was built to withstand the pressure of pack ice. In 1893-6, Nansen attempted to reach the North Pole by using the natural drift of the polar ice in the prevailing east–west current. But the ice moved unpredictably, thwarting the Norwegian's plan.*

In 330 BC a Greek navigator named Pytheas sailed as far as Shetland, or possibly even Iceland, in the first recorded exploration north toward the Arctic. Thereafter, up to the end of the 18th century, a succession of northern European sailors – the Vikings, the Dutch and the Danes – explored and named territories around the Arctic Circle: Iceland, Greenland, Novaya Zemlya, the Aleutian Islands, the Lancaster and Bering Straits.

In 1893 the Norwegian explorer Fridtjof Nansen, using a team of sled-dogs over the Arctic ice, came within 450km of the North Pole, the elusive location that marks the point through which the Earth's rotational axis passes. Four years later Salomon Andrée, a Swede, was killed attempting a balloon flight over the Pole, while in 1899 Louis Amedée de Savoie, the duke of Abruzzo, advanced to within 400km of the Pole. On a voyage lasting from 1903 to 1906, the Norwegian Roald Amundsen became the first explorer to find a way through the Northwest Passage, a long-sought route from the Atlantic to the Pacific Ocean through the ice-bound islands to the north of Canada.

Because the North Pole is located on ice above the Arctic Ocean, not on solid land, it can be difficult to verify claims to have reached it. Two Americans, Robert Peary and Frederick A Cook, made rival claims to have been the first. Peary claimed to have planted the Star-Spangled Banner at the North Pole on 6 April, 1909, while Cook maintained he

DISPUTED TERRITORY

The Arctic region comprises the Arctic Ocean – a large part of which stays frozen as pack ice all year round – and the northernmost areas of the countries surrounding it: the Scandinavian peninsula, the north of European Russia (the Kola Peninsula) and Siberia, the north of Alaska and Canada, Greenland and Spitsbergen. The Arctic Ocean is governed by the 1982 UN Convention on the Law of the Sea, although this was not ratified by the USA. The Arctic sea floor is rich in natural gas, oil and other minerals. In 2007 the manned mini-submersibles *Mir-1* and *Mir-2* planted a rust-proof titanium alloy flag on the sea bed, 4,261m beneath the ocean's surface at the geographic North Pole, staking Rusia's claim to the region's resources.

performed the feat a year earlier, on 21 April 1908. After studying the logbooks and charts of both explorers, the US Congress came down in favour of Peary, but the prevailing view today is that neither man reached the Pole.

The first verified flight over the North Pole took place on 12 May, 1926, by the Italian airship *Norge*. On board were Amundsen, the pilot and designer of the airship Umberto Nobile and the American explorer Lincoln Ellsworth. In 1948 an expedition from the Soviet Union became the first to set foot at the geographic North Pole, reaching it from an aircraft that landed on the ice. They were followed in 1959 by sailors from the American nuclear-powered submarine USS *Skate*, which surfaced at the Pole.

The South Pole – the final frontier

The Antarctic first appeared on maps in the 16th century, marked as *Terra incognita australis* – 'the unknown southern land'. Shown as being about the size of Europe, the southern land had long been rumoured to be rich in treasure. Captain James Cook was the first to try to find out what was really there, sailing his ship into the Antarctic Circle and reaching a southern latitude of 66°33' on 17 January, 1773. But Cook's expedition did not reach the coast of Antarctica. The discovery

Polar pioneer
French ethnologist and explorer Paul-Émile Victor (right, 1907–95) looks out over the Arctic ice on one of his 25 polar expeditions. In the 1930s he led expeditions to Greenland and lived among its indigenous communities.

Under threat
The traditional way of life of the Inuit (below), dating back some 10,000 years, changed dramatically following contact with Europeans and the advent of air travel.

At home on ice
Adélie penguins (Pygoscelis adeliae, *right) are the most common species of penguin in the Antarctic. They were named in honour of the wife of the 19th-century French explorer Jules Dumont d'Urville. The men of Ernest Shackleton's ill-fated expedition survived by catching these birds to eat when their food supplies ran out.*

of the continent, in around 1820, is variously attributed to the English sea captain Edward Bransfield, the American Nathaniel Palmer, or the Russian Baron Fabian Gottlieb von Bellingshausen. Thereafter, the sixth continent was largely forgotten until 1895, when the International Geographical Congress, held in London, announced that 'the exploration of the Antarctic region is the greatest piece of geographical exploration still to be undertaken'. A series of expeditions and achievements then came thick and fast.

An expedition led by the Belgian Adrien de Gerlache de Gomery from March 1897 to 1899 was the first to overwinter on the Antarctic ice. The first permanent base was established by the Norwegian Carsten Egeberg Borchgrevink at Cape Adare. National claims to parts of Antarctica as overseas territories began in 1908 with Britain's declaration of sovereignty over a sector encompassing the Weddell Sea. The seasoned Arctic explorer Roald Amundsen won the race to the South Pole, beating a British team led by Robert Falcon Scott. Amundsen reached his goal on 14 December, 1911, then returned safely home. Scott's party arrived on 17 January, 1912, and perished on the return journey. Perhaps the most remarkable story of Antarctic exploration is that of the Imperial Trans-Antarctic Expedition of 1914–17, led by Ernest Shackleton. He failed in his bid to cross the continent from sea to sea via the pole after his ship was crushed by ice, but against all the odds Shackleton succeeded in reaching South Georgia to get help and saved his entire team.

Except for the first overflight of the South Pole in 1929, little exploration of Antarctica took place in the interwar period. In 1946 a flotilla of US Navy ships commanded by Admiral Richard Byrd, comprising an aircraft carrier, icebreakers and submarines with some 4,000 men, surveyed and mapped a large section of the Antarctic coast. The following year, the *Expeditions Polaires Françaises* (EPF) of Paul-Émile Victor planted the French flag on Adélie Land, discovered by Jules Dumont d'Urville a century earlier. A joint Norwegian, British and Swedish expedition of 1949–52 gathered extensive scientific information on the region. In 1954 the Australians built a permanent base at Mawson Station, at the foot of the Lambert Glacier.

A FROZEN DESERT

Antarctica has a total ground area of 14 million km^2. It is covered by an ice sheet with an average thickness of 2.2km, which accounts for 90 per cent of all the permanent ice on Earth. Antarctica is unique as a continent in that it belongs to nobody; it is also the driest, coldest and windiest place on the planet. It receives no solar radiation (sunlight) during the winter, a period lasting from between a few days on the Antarctic coast to almost six months at the South Pole. The continent has one active volcano, Mount Erebus, at 3,794m; the highest peak on the continent is the Vinson Massif at 4,897m. There are more than 40 research stations where scientists overwinter, maintained by 17 countries. These are mostly located on the coast.

Preparing the way
American Richard E Byrd established the permanent base of 'Little America' during his first expedition to the South Pole in 1928–30. It was from there that he took off to make the first flight over the Pole itself in November 1929. Before the flight, a dog sled team (above) was sent out to leave caches of food and fuel at intervals along the route in case of a forced landing. The enterprise involved 50 men, 650 tons of supplies and 95 huskies.

A momentous year

The research carried out by various nations on Antarctica bore fruit in 1957, which was proclaimed the International Geophysical Year (IGY). Sixty-seven countries agreed to pool resources and collaborate on a systematic

Home of scientific research

Entombed in ice for 35 million years, the Antarctic has now become a favoured spot for scientific research. Since 1962 the British Antarctic Survey (BAS), with headquarters in Cambridge, has undertaken scientific research on and around the continent, working with scientists from over 30 countries. It has three stations in the Antarctic – at Rothera, Halley and Signy – and two on South Georgia at King Edward Point and Bird Island. In 1985 the BAS drew the world's attention to the hole in the ozone layer, and they were also first to link the rapid warming of the Antarctic Peninsula with human activity. In addition, they have revealed the amazing survival stories of creatures such as the kittiwake adapted to living in the extreme cold. Operations in the Antarctic are carried out by ships strengthened to withstand the power of ice and equipped with the most up-to-date facilities for oceanographic research.

study of the planet, which ran until December 1958. Antarctica was the main focus of the IGY, with research being carried out at the 48 stations established on the continent. These were mostly situated around the coast, though four were on the ice shelf in the interior, the most inhospitable terrain on Earth. One practical outcome of this international cooperation was the Antarctic Treaty, which was signed in Washington DC in 1959 – a remarkable achievement at the height of the Cold War. The agreement came into force on 23 June, 1961, and established Antarctica as a scientific preserve, free from all military or nuclear activity; to date, 47 countries have signed and ratified the treaty. It was broadened in 1998 by the Madrid Protocol, which designated the entire area south of the 60th Parallel as a nature reserve, in which all mineral exploitation is banned until at least 2041 (a provision that can be indefinitely renewed).

Working together
The two main blocks of the Concordia Station on Antarctica comprise living quarters and a research facility. They are covered with panels of highly efficient insulation material, which makes it possible to live on the surface of an icecap where air temperatures often fall to –50°C and beyond.

Other international research is carried out by installations such as the Concordia Station, a joint French and Italian venture established in 2005 at a location known as Dome C. This all-year facility lies 1,100km inland at an elevation of 3,233 metres. Glaciologists, have been conducting ice-core drilling at Dome C since the 1970s. Bubbles of air trapped within columns of ice extracted from deep within the Antarctic ice sheet enable the scientists to

Global warming in ice
Recent analysis of ice-core samples from the Antarctic have shown that the CO_2 content of the Earth's atmosphere is now greater than it has been at any time in the past 650,000 years.

Unique creatures
Right: A crocodile icefish, an Antarctic species whose blood is kept fluid by the absence of haemoglobin. This adaptation to the extreme cold of the Southern Ocean makes the species vulnerable to rising sea temperatures.

Ice cliff
Far right: Ice crashes into the sea from the Hubbard Glacier in Alaska, the largest tidewater glacier in North America. Each year, 20,000 tonnes of ice break off from polar ice caps and glaciers, forming icebergs. As they melt, they raise the global sea level.

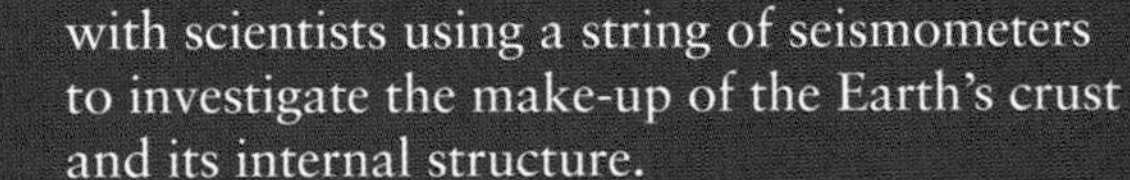

reconstruct the composition of the Earth's atmosphere as it was thousands of years ago. By measuring the content of harmful greenhouse gases such as carbon dioxide, methane and nitrous oxide and comparing it with today's levels, they have built up a picture of how the Earth's climate is evolving. Currently, researchers are able to go as far back as 800,000 years. One of Concordia's principal purposes is to investigate the feasibility of constructing an observatory at the site, as the unpolluted Antarctic air would allow observation of the stars even when the Sun is at a high angle of elevation. Geophysical research is also carried out here, with scientists using a string of seismometers to investigate the make-up of the Earth's crust and its internal structure.

Other scientists working in Antarctica include oceanographers, who examine the marine ecosystem of the Southern Ocean; geologists, who collect micrometeorites; and biologists, who study some of the continent's extraordinary fauna. These include the crocodile icefish (*Channichthyidae* sp.), whose blood contains no haemoglobin, making it transparent, and cod-like icefish (*Notothenia*), which have adapted to the intense cold by

THE DANGERS OF TOURISM

Tourism to the Antarctic is a relatively recent phenomenon, having begun in the 1950s, when Argentina and Chile allowed more than 500 paying passengers to travel to the South Shetland Islands. In the 1960s the concept of Antarctic cruises became a reality with the launching of a purpose-built ship, the *Lindblad Explorer*. In the early 1990s the number of people visiting the region equalled that of the scientists and technicians working on the continent for the first time. Since 2000 there has been a marked increase in extreme sports holidays in the regions, as well as other tourism. It is feared that unchecked growth in visitor numbers will disturb animals and birds during their breeding season, damage vegetation trampled underfoot and bring an increased risk of pollution from shipping accidents.

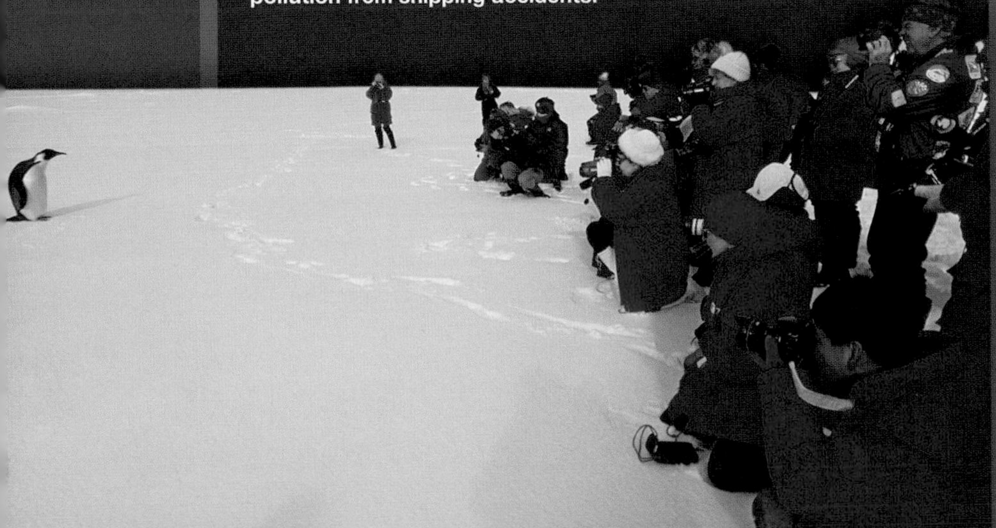

MYSTERY UNDER THE ICE

In 1973 radar imaging revealed the presence of a vast lake beneath the Antarctic ice sheet, directly below the Vostok research station in the east of the continent. The largest by far of some 140 subglacial Antarctic lakes, Lake Vostok has a surface area of around 14,000km^2 (roughly the size of Lake Ontario in the Great Lakes); it is 224km long and several hundred metres deep. Scientists believe that its isolation from the external environment for millions of years may have given rise to unique primitive bacterial life forms.

THE BIG MELT
The melting of the Arctic ice cap is accelerating under the influence of global warming. In 2007 satellite images, first available from 1978, showed that since the 1980s the surface area of the ice has shrunk by some 2 million km².

generating antifreeze proteins in their blood. Physiologists in Antarctica are monitoring human ability to adapt psychologically and socially to extreme environments. For example, the European Space Agency maintains a presence at the Concordia Station, preparing astronauts for long-term space missions. And chemists are studying the hole in the ozone layer over Antarctica, which reaches its largest extent in October each year: its maximum size to date has been around 27 million km². Caused by the increasing presence of pollutants, especially chlorofluorocarbons (CFCs) in the atmosphere, it is a cause of great concern as the ozone layer is what shields the Earth from harmful ultra-violet radiation.

Feats of endurance

In 1968–9 British explorer Wally Herbert and three companions became the first to make a surface crossing of the Arctic Ocean by its longest axis, through the North Pole, with the aid of dog sleds. On 11 May, 1986, the French explorer Jean-Louis Étienne completed a solo walk to the North Pole, an epic journey that lasted 63 days. In 1993 Ranulph Fiennes and Mike Stroud completed the first unsupported walk across the Antarctic landmass, dragging all their equipment on sleds behind them.

The Arctic pack ice has been retreating in recent decades, most drastically in summer months, raising fears about catastrophic coastal flooding in many parts of the world if nothing is done to combat global warming. Scientists of all disciplines are concerned to safeguard the harsh, fragile yet hugely rich ecosystems of both the Arctic and Antarctic.

Threatened species
Global warming poses a direct threat to Arctic species such as the polar bear, as melting of the polar ice sheet drastically reduces their fishing opportunities.

A new era for television

The invention of the videorecorder in 1954 freed television from the shackles of live broadcasting. The first video tape recorders were designed to help broadcasters to organise their programmes and schedules more efficiently. Gradually, the technology became available to all, changing the way that people interact with TV.

By 1953, some 7 million American homes had a television. All TV broadcasts at the time were live, which presented a major problem for the national networks, such as ABC and CBS, because the time difference across the country meant they could not broadcast the same programmes on the east and west coasts. This was where the videorecorder came in.

Following a demonstration in 1951 by Mincom, a subsidiary of the 3-M magnetic tape company, the recording giant RCA began to develop a black-and-white videotape recorder, which it launched three years later. The images were recorded horizontally on a 2-inch (2cm) wide magnetic tape, much as on an audiotape recorder. But because television images contained some 300 to 500 times more information than audio, the tape had to run at the much faster speed of 9m/sec, requiring more complex and durable recording heads.

Time-shift programming

In 1956, the American Ampex company came up with a more practical solution in the form of its VR1000 machine, which recorded video images vertically onto the 2-inch tape. This recording medium was called 'quadruplex', from the four magnetic record/reproduce heads mounted on a headwheel that spun transversely (that is, width-wise) across the tape at a speed of almost 15,000rpm. This enabled the tape itself to run much slower, at 0.38m/sec. On 30 November, 1956, the first time-shift programme, recorded in New York on a VR1000, was broadcast in Hollywood, Los Angeles, three hours later.

Ampex was also responsible for introducing the first colour VTR, in 1958. The following year, the Japanese Hitachi company made the process simpler and more cost-effective through the introduction of a single spinning recording head that recorded video obliquely onto tape ('helical scan'). Philips of the Netherlands refined this system in 1964, adding a second head to achieve a more concentrated signal. All subsequent video recorders adopted this basic principle.

Dead end

The Sony Betamax VCR (above) came on the market in 1975. Hollywood studios like Universal and Walt Disney fiercely opposed home recording. In 1984 a landmark verdict in favour of Sony ruled that all home taping of TV broadcasts or films was legal. Yet when home recording began to boom in the 1980s, Sony's Betamax was eclipsed by the rival VHS, developed by JVC.

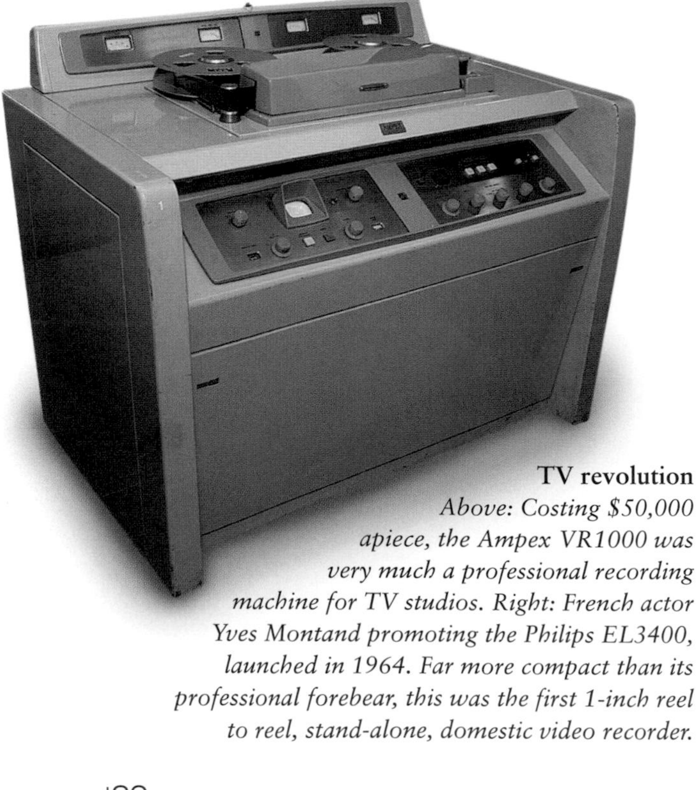

TV revolution

Above: Costing $50,000 apiece, the Ampex VR1000 was very much a professional recording machine for TV studios. Right: French actor Yves Montand promoting the Philips EL3400, launched in 1964. Far more compact than its professional forebear, this was the first 1-inch reel to reel, stand-alone, domestic video recorder.

RISING RECORDING TIMES

Over 25 years, the recording time available on a standard half-inch wide VCR cassette tape increased exponentially:

- 1972 Philips VCR – 1 hr
- 1975 Sony Betamax – 1 hr, 2 hrs, then 3 hrs
- 1975 JVC VHS – 3 hrs, 4 hrs, then 5 hrs
- 1998 JVC Digital VHS – 2 to 40 hrs

In 1967 Ampex launched the VR3000, the first portable video recorder enabling recordings to be made outdoors. The next innovation came in 1971 when Sony introduced the U-Matic; this machine, which used three-quarter-inch wide tape, replaced the open-reel mechanism with a cassette format. Teamed with a shoulder-held video camera, the apparatus gave journalists the freedom to do location reports without the need for a full outside broadcast unit. By 1982, the Betacam, a superior videocamera system developed jointly by Sony and Thomson, became the industry standard. This was supplanted in turn in 1993 by the digital Betacam.

Video rental
The 2008 film 'Be Kind Rewind' tells the surreal story of two video rental store employees who accidentally wipe all the tapes. Video stores were popular from the 1980s onwards, but have recently faced competition from Internet sites that post out DVDs and Blu-ray discs to viewers in return for a small monthly fee.

Young consumers
Home video players had a major impact on the children's movie market, as families bought tapes of favourite films to watch at home.

The format war

The VR1000 was huge and heavy: 20 years were to pass before the videorecorder became a genuinely mass-market appliance. In the 1960s several manufacturers, including Sony, Philips, Sears, RCA and Motorola, tried to market open-reel VTRs, but the equipment was expensive, unwieldy and unreliable, and failed to make an impact.

The real breakthrough in the home market came in 1972, when Philips introduced its Video Cassette Recording System. Equipped with its own timer and tuner, and using a tape cassette with an hour-long recording time, the VCRS could tape programmes directly from the television and set the pattern for all models thereafter. Sony followed suit in 1975 with its own Betamax system, a development of the U-Matic, also offering one hour of recording time. In October that same year, the newcomer JVC arrived on the scene with its VHS system.

As home recording of TV programmes took off, movie distributors began to release pre-recorded feature films on cassette. In a scramble to corner this lucrative new market, a war broke out between the various VCR formats, which were incompatible with one another. While Sony's Betamax was the clear leader in 1981, five years later VHS had captured the lion's share of the market. The relatively low cost of the VHS cassettes, which sacrificed quality to achieve 3 hours' running time, helped it to win out over the Sony and Philips systems. (Some commentators point to the fact that the makers of blue movies plumped for VHS as a factor in JVC's victory.)

Nowadays, viewers are faced with a widening range of options, including DVDs, Blu-ray discs, digital video recorders and VOD (video on demand) through set-top boxes. But what sparked the home entertainment revolution was the analogue VCR.

NEW OPPORTUNITY

At first, the film industry saw home-taping on video as a threat to box-office takings in cinemas. But they soon came to realise that extra revenue could be generated from releasing movies on VHS cassettes, and later DVD. Home viewing now represents such an important share of the market that some video producers even help to finance film production.

Electricity from nuclear power

There had already been ample evidence of the awesome power of nuclear energy in the atomic bomb. In the post-war period, atomic scientists turned their minds to the civil use of nuclear power. From the mid-1950s onwards, nuclear power stations began to come on stream in the Soviet Union, the USA and Western Europe

With its long, straight avenues lined with identical apartment blocks, Obninsk could have been any one of the hundreds of new towns built in the Soviet Union in the 1950s, from the Ukraine to the remotest reaches of Siberia. But here, around 100km southwest of Moscow, the inhabitants were all privileged members of the Soviet scientific élite. Obninsk was a research city where everything was geared to a single goal – the development of civil nuclear power in the USSR. In June 1954, in the presence of a delegation of dignitaries from the Moscow Politburo, the city enjoyed its proudest moment as the AM-1 nuclear reactor (AM stood for *atom mirny*, literally 'peaceful atom') produced its first megawatts of energy. The energy released by splitting the atom was being exploited not to make bombs but to generate electricity. After three weeks of intensive tests, AM-1 was connected to the country's national grid. It would produce power for almost half a century, finally being decommissioned in 2002. Just nine years before AM-1 came on stream, the Japanese cities of Hiroshima and Nagasaki had been flattened by the first atomic bombs. Now this same energy, released by the fission of atomic nuclei, was being deployed for peaceful ends.

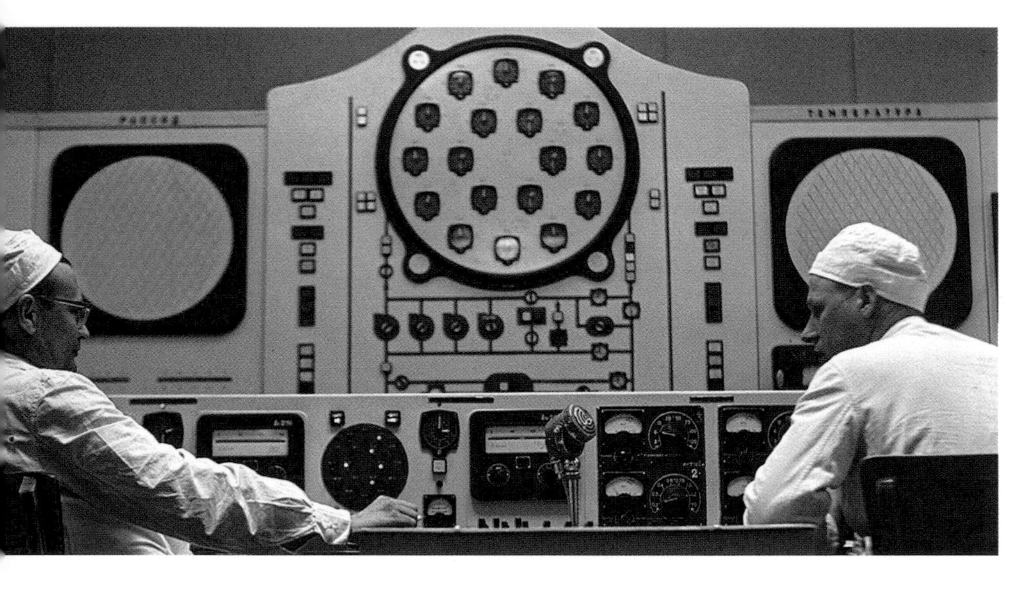

Prestige project *Workers at the Obninsk nuclear plant in Russia in 1964. Operating this pioneering reactor cost 30 million roubles a year, but its output only recouped 10 per cent of this figure.*

Enrico Fermi's 'pile'

Great strides had been made in nuclear physics by the 150,000 or so scientists, engineers and technicians who worked on the Manhattan Project, the US programme to build the atomic bomb. In 1942 one of the principal physicists on the project, Enrico Fermi, built the world's first nuclear reactor on a rackets court at the University of Chicago. It consisted of a 'core' of uranium pellets, separated from one another by graphite blocks to slow down the neutrons. In this slowed state, they stood a better chance of colliding with uranium atoms.

Fermi described his reactor as a 'crude pile of black bricks and wooden timbers', hence its official name of Chicago Pile-1 (CP-1). All nuclear reactors henceforth were referred to as 'atomic piles'. Cadmium-coated rods that absorbed neutrons prevented the reaction from running out of control.

Ulterior motives

Although the Manhattan Project was wound down at the end of the Second World War, it gave rise to an entirely new nuclear industry, involving close collaboration among a complex network of engineering firms, boilermakers, chemical companies and others. Naturally, the United States had a head start in this field, further stimulated by the formation of the Atomic Energy Commission (AEC). The aim of the AEC was to develop nuclear power as a source of energy for domestic and business use. The idea was to create a viable reactor as the core of a traditional power station, in which the nuclear heat source would turn water to vapour, thereby driving a turbine linked to a generator.

At 13.50 hours on 20 December, 1951, the Experimental Breeder Reactor at the Department of Energy's Idaho National Engineering and Environmental Laboratory generated the world's first electricity from nuclear fission, lighting four 200-watt bulbs. Unlike the contemporary Soviet programme, though, EBR-1 was only intended as an experimental facility.

French fission *Replacing fuel rods at the Marcoule nuclear power station in 1958. This site, France's first nuclear facility, was decommissioned in 1997, and now houses a research centre specialising in the reprocessing of high-level radioactive waste.*

Business as usual *Despite the serious accident at Chernobyl, the three undamaged reactors kept operating until 2000 because of the high demand for electricity. Here, technicians examine Chernobyl's Reactor Number 1, which was decommissioned in 1996.*

Other technologically advanced nations rushed to join the nuclear club. British scientists developed a nuclear power plant cooled by carbon dioxide gas at Calder Hall in Cumbria. The type of reactors installed there were officially known as 'Magnox', from the magnesium-oxide alloy used to coat their fuel rods. Like Fermi's first pile, these reactors were graphite-moderated, and used natural uranium as fuel. Calder Hall was opened by Queen Elizabeth II on 17 October, 1956; its operators claimed the four 50MW reactors would generate 'electricity too cheap to meter', but another purpose of the facility was to produce weapons-grade plutonium-239 for Britain's atomic-bomb programme. The same motivation lay behind the construction of France's first nuclear power station at Marcoule in the south of the country. Built to a very similar design to the British Magnox, the Marcoule reactors had an output of 43MW each; over their lifetime they supplied 11 billion kWh to the French national grid. Germany, meanwhile, embarked on a strictly peaceful nuclear power programme in 1961.

The use of natural uranium in these early plants posed a major problem. Since only around 0.7 per cent of this mineral is made up of uranium-135 (the only fissile isotope), huge quantities were required to run the reactors. The USSR got round this difficulty with their new RBMK reactor design, which used

THE ENDURING LEGACY OF CHERNOBYL

The world's worst accident at a nuclear power station began to unfold at Chernobyl, near the Ukrainian capital Kiev, on 26 April, 1986. During an authorised systems test – the first of a chain of human errors – the plant's Reactor Number 4 ran out of control, rupturing the reactor vessel and causing a series of explosions. One major safety flaw in RBMK plants like Chernobyl was the lack of a hardened-concrete containment building; the resulting fire in the graphite moderator block sent a plume of radioactive smoke spewing out over Europe. Around 350,000 people living within a 30-km radius of the station were evacuated. More than 20 years after the incident, the area around Chernobyl remains an exclusion zone. Around 50 of the workers and firemen involved in fighting the blaze died of severe radiation exposure, both during the immediate aftermath and in the years that followed. The World Health Organization estimates that some 4,000 people will eventually die of thyroid cancers and other diseases brought on by the accident. A total of 5.6 million people in Europe were affected with larger or smaller doses of radiation and scientists are still monitoring the long-term effects. The true human cost, particularly those of inherited defects, may never be fully known.

NUCLEAR FUEL PROCESSING

Uranium ores, the source of all fuel for nuclear power stations, occur mainly in Canada, Australia, Kazakhstan and Niger. As they contain only a tiny fraction of uranium (1–5kg per tonne), primary processing involves a series of physical and chemical operations to extract the viable part of these ores. Facilities near uranium mines produce a powder called 'yellowcake', containing 75 per cent uranium. But natural uranium comprises two separate atoms: uranium-238 (99.3 per cent) and uranium-235 (0.7 per cent, the sole radioactive isotope). This means that the yellowcake must be enriched for it to become nuclear fuel. Isotope separation techniques such as gas diffusion and ultracentrifugation increases the radioactive composition of U-235 to around 4 per cent for reactor-grade fuel. Gas diffusion entails passing the element, in the gaseous form of uranium hexafluoride, through a series of ultra-fine membranes with millions of tiny holes. The lighter U-235 atoms travel faster and diffuse more readily through the membranes, thus eventually producing the desired degree of enrichment.

Ultracentrifugation involves placing the yellowcake in high-speed centrifuges, which draw the heavier U-238 to the outer wall, while the U-235 molecules collect at the centre. Finally, the enriched end-product is made into uranium oxide fuel pellets, which are packed into fuel rods and lowered into the reactor core. This form of fuel is known as UOX, while an alternative fuel, MOX (mixed oxide), is made from a compound of 7 per cent plutonium and depleted uranium.

Fuel of the future?
When pellets made from the natural metallic element thorium are bombarded with neutrons, they transform into U-233, a fissile artificial isotope of uranium.

Opencast pit
Excavation of uranium ore at the vast Key Lake opencast facility in Saskatchewan, Canada (above). Once the world's largest uranium deposit, operations ceased here in 1997, after 14 years of mining, and the site has been converted into a nuclear waste treatment plant.

enriched uranium. Yet these water-cooled reactors proved difficult to operate and were potentially dangerous, a fact brought home by the Chernobyl disaster in 1986.

PWR reactors

In the event, the technology that ended up being adopted worldwide was the pressurised water reactor (PWR), a type first devised in the United States in the late 1940s. It began life as the propulsion unit for the world's first nuclear-powered submarine, the USS *Nautilus* (1955). The fuel used in a PWR is enriched uranium, with pressurised water as the neutron moderator. In the primary circuit, light water acts as a

Eerie glow
Pressurised water reactors are surrounded by deactivation pools, where radioactive components are placed in order to allow the heat generated in them by the nuclear reaction in the reactor core to escape. The blue colour of these pools arises from a phenomenon known as 'Cherenkov radiation,' which occurs as a result of charged radioactive particles moving faster than the speed of light in the water.

coolant to reduce the heat generated by the reactor core. In the process, some of this superheated water turns into steam, which, after being fed into the secondary circuit, drives the turbines to produce electricity. It is then converted back to liquid in a condenser fed constantly by cold water drawn from a river.

The first power station of this type, Shippingport in Pennsylvania, came on stream in 1957. Many countries followed America's lead, replacing earlier reactor types with PWRs. Almost all of France's 58 nuclear power stations, for example, are PWRs, while Germany has 14. In total, some 60 per cent of all nuclear-generated electricity worldwide

THEN AND NOW – THE STORY OF DOUNREAY

The experimental fast breeder reactor at Dounreay in the north of Scotland, nine miles west of Thurso, was built between 1955 and 1958 and led British research and development in the field of nuclear energy during the 1950s and 60s. Housed inside a steel sphere, it became the first fast reactor in the world to provide electricity to a national grid in 1961. The reactor closed down in 1977, but this is far from the end of the story, as decommissioning presents a significant challenge. A blanket of uranium and plutonium breeder material is submerged in 57 tonnes of liquid metal. Both need to be removed to enable the rest of the reactor to be cleaned out and dismantled. The future of the iconic sphere itself, which is recognised as a local landmark and as a national symbol of atomic heritage, is also under debate.

A LONG-TERM PROBLEM

The management of high-level radioactive waste (HLW), which results from the reprocessing of spent nuclear fuel, is one of the major headaches facing the nuclear industry. HLW accounts for more than 95 per cent of all the radioactivity produced in the nuclear power process. It remains dangerous for at least 30 years – in some cases, for centuries – as a result of the presence within it of the fission products caesium-134, caesium-137 and strontium-90. At present, storage of such waste above ground at nuclear sites involves vitrifying it (making it stable) and sealing it in stainless-steel containers. But as the amount of HLW increases, the proposed long-term solution is to bury it at a depth of at least 500 metres in seams of impermeable clay, so-called 'deep geologic disposal'. Yet locating potential sites and gaining planning permission for such repositories in the face of local opposition is far from straightforward.

Action against radioactivity
Dressed as drums of radioactive waste, anti-nuclear demonstrators from Greenpeace protest outside a power station. Public opposition to nuclear power was galvanised by the accidents at Three Mile Island and Chernobyl.

Remote control
Left: at a nuclear fuel reprocessing plant, mechanical arms are used to load highly toxic nuclear waste into radiation-proof containers.

comes from such plants. Britain, exceptionally, stuck with the advanced gas-cooled reactor (AGR) design; only one of its 16 nuclear power stations – Sizewell B in Suffolk – is a PWR.

The Middle East oil crisis of 1973 boosted the allure of nuclear power. To ensure future energy security, most industrialised nations responded to the crisis by bringing forward their civil nuclear power programmes. The economic argument seemed conclusive: 1kg of uranium can generate as much electricity as 2,500 tonnes of coal. As early as 1967, the USA had begun building 50 new nuclear power stations.

Growing opposition

But in the late 1970s the image of nuclear power became seriously tarnished. Enduring concerns about the storage of toxic waste were compounded by a serious accident – a partial core meltdown – at the US power plant at Three Mile Island in March 1979. Thereafter, a significant section of public opinion turned against this form of energy. Plans to construct

FUSION POWER

Nuclear fusion – the reaction that occurs at the core of stars and the principle behind the hydrogen bomb – involves fusing together two light atomic nuclei, such as hydrogen and helium, to form a heavier nucleus. In the process, large amounts of energy are released. Because fusion can be achieved with abundant fuels such as tritium and lithium and produces no waste, it has been hailed as the successor to nuclear power. But the technological difficulties are immense. Primarily, atomic nuclei, which repel one another like magnets, can only be fused by subjecting them to the kind of pressure that exists at the heart of the Sun. Also, there is the problem of how to contain plasma at temperatures of tens of millions of degrees. To date, the largest fusion experiment has been the Joint European Torus (JET) project at the Culham Laboratory near Abingdon in Oxfordshire.

Nuclear nation

Cooling towers at the nuclear power station at Tricastin in southeastern France. This country placed great emphasis on nuclear power, building 59 plants that generate three-quarters of its electricity requirements. This is the highest percentage anywhere in the world.

new stations were met with fierce, sometimes violent, opposition. Some countries, such as Sweden and Italy, abandoned their civil nuclear programmes altogether, while others scaled back development. Others, notably France, the USSR and Japan, continued full-steam ahead.

Since the late 1990s, with growing alarm over global warming caused by burning fossil fuels, the nuclear power industry has enjoyed something of a revival. Unlike coal and gas-fired power stations, nuclear plants emit no greenhouse gases as by-products. Plus, as fossil fuels dwindle, creating an energy gap that cannot be filled by renewables like solar, wave and wind power, nuclear power will become a vital piece of the jigsaw in meeting energy needs. China, India and many other rapidly industrialising nations have ongoing civil nuclear programmes. As of 2009, there were 52 nuclear reactors under construction around the world, in addition to the 436 already in operation. Yet for all this, nuclear power still only accounts for 17 per cent of the total energy generated worldwide, outstripped even by the 19 per cent generated from renewable resources – and far behind the 64 per cent still accounted for by fossil fuels such as oil, natural gas and coal.

Non-stick pans 1954

Carefree cooking
An advertisement for Tefal pans from the 1960s (right). So widespread did Teflon become that it even began to be used to refer to people to whom scandal did not 'stick'. At one time commentators began calling former prime minister Tony Blair 'Teflon Tony'.

In 1954 French engineer Marc Grégoire was employed in the aeronautics industry. He was investigating the remarkable properties of Teflon, a material notable for its chemical inertness and the frictionless surface it produced. On one of his holidays, while trying to use Teflon's smoothness to improve the performance of his fishing tackle, his wife made the casual suggestion that he should coat their cooking pans with it to make them non-stick.

Teflon, or to give it its proper chemical name polytetrafluoroethylene (PTFE), had been invented in 1938 by the American chemist Roy Plunkett during research into refrigeration; the DuPont Chemical Company marketed it under its familiar brand name from 1949. One extraordinary property of this synthetic polymer is that it only breaks down at high temperatures (over 260°C) and has a very high melting point (327°C). In contrast, cooking oils burn at around 180°C. But the absolute smoothness of Teflon, one of its main advantages, was also a major drawback: Grégoire could not get it to adhere to the metal bases of the pans. Finally, he hit on the idea of keying the surface of the aluminium with an acid solution before coating it with Teflon powder. The acid created tiny bumps on the surface of the metal, which the Teflon powder clung to forming a smooth surface.

SAFETY CONCERNS

Rumours that Teflon might be harmful have never been substantiated in tests. Even so, DuPont is planning to phase out production by 2015. The US Environmental Protection Agency requires all companies to limit public exposure to perfluorooctanoic acid (PFOA), a constituent ingredient of Teflon, which is thought to be mildly carcinogenic.

Teflon takes off

Grégoire patented his invention in 1954 and two years later formed a company to make and market his pans, the Société Tefal (an amalgam of the words 'Teflon' and 'aluminum'). In demonstrations at stores and trade fairs, he promoted his product as 'the saucepan that truly never sticks'. Yet business remained sluggish until a photo appeared in 1961 of Jackie Kennedy, the US First Lady, holding a Tefal pan. Sales instantly soared, with orders coming in of nearly 5,000 pans a week. By 1968 Tefal had become the leading manufacturer of kitchen products in France.

Today Teflon is used on a wide range of appliances, including casseroles, toasters, deep fryers, pressure cookers and even steam irons. Yet the fact that the surface always eventually wears out prompted the launch of rival coatings such as ceramic titanium and more recently nanodiamonds. The extreme durability of this surface means that, in theory, a non-stick pan could last for ever.

Polio vaccine 1954

Polio pioneer
Jonas Salk injects a girl with his polio vaccine in 1954. Nowadays, children are usually given an anti-polio jab in conjunction with other vaccines to prevent diseases such as tetanus, diphtheria and whooping-cough.

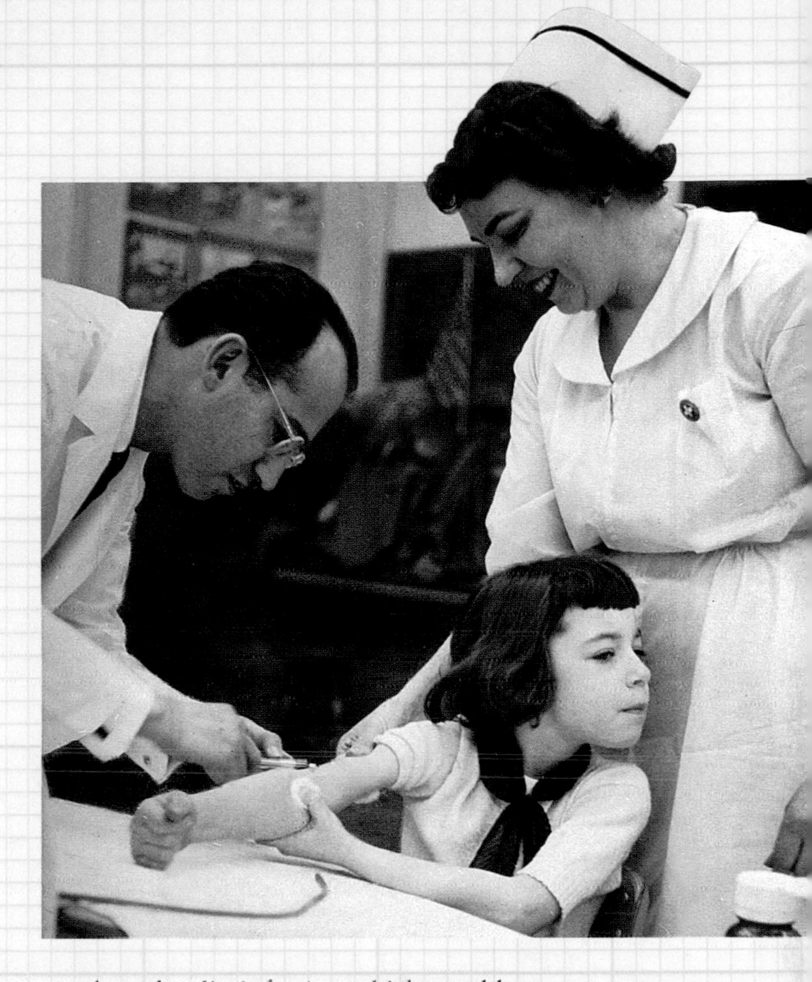

Global scourge
Sudanese children are inoculated against polio at a refugee camp in Chad in 2004. Despite efforts to eradicate the disease, it remains endemic in many parts of the world.

In 1952 a polio epidemic struck the USA, affecting more than 47,000 people. Two years later, on 26 April, 1954, a young medical researcher named Jonas E Salk developed a vaccine that he used to inoculate almost 2 million children against polio. The success of this clinical trial was soon apparent: up to 80 per cent of those who were given the inoculation did not contract the terrible disease. In the most severe cases poliomyelitis, which is transmitted by consuming contaminated food or water, caused total paralysis and often death by respiratory failure. Many who recovered suffered permanent damage to the central nervous system. The search for an effective vaccine had been going on in the United States since the 1930s.

Dead or alive?

Following a classic procedure pioneered by Edward Jenner and Louis Pasteur, some researchers, including the eminent American virologist Albert B Sabin, sought a vaccine based on an attenuated (weakened) form of the live virus. The aim was to give the patient a weakened polio infection which would not develop the worst symptoms of the disease. Others, including Jonas Salk, worked on producing an inactive vaccine using a 'killed' virus to stimulate the production of antibodies without inducing the disease at all.

The two research camps came up with vaccines at around the same time, both of which required regular boosters. Sabin's was taken orally on a sugar lump – unlike Salk's vaccine, which was injected – and proved ideal for mass vaccination programmes in countries where the disease was rife. But it did have one serious drawback: even if only in a very small number of cases, the attenuated virus had the potential to revert to a form that could paralyse. Over time, both vaccines have been improved and manufactured on an industrial scale. The incidence of polio in industrialised countries fell dramatically after mass vaccination programmes in the 1950s. In 1988 the World Health Organization and UNICEF launched a campaign to eradicate polio worldwide, which has been very successful.

THE SLOW MARCH TO A POLIO-FREE WORLD

When the World Health Organization and UNICEF began a campaign to eradicate polio around the globe in the late 1980s, the disease was present in 100 countries, affecting some 350,000 people annually. By 2001, the programme had reduced infection to just 483 reported cases. Yet the virus persists in pockets of Asia and Africa, causing periodic surges in cases if countries stop a vaccination programme prematurely. In 2008 around 1,500 cases were recorded, more than half of them in Nigeria.

A life-saving operation

In 1954 the world's first successful organ transplant was performed by an American medical team. One of the biggest problems, which took many years to overcome, was that of organ rejection by the patient's own body. Transplantation is common today, but difficulties remain with a shortage of donors and complex questions of medical ethics.

'Organ transplantation, a simple surgical curiosity today, may one day have a definite practical interest.' So wrote the renowned French biologist Alexis Carrel in 1902. Six years later, this visionary surgeon – who did pioneering work on vascular sutures – succeeded, in an experimental procedure, in transplanting a dog's kidney to the vessels in its neck; the dog lived for 17 months. But doctors were still far from being able to perform fully successful organ transplants. Over the next 50 years several transplant attempts were made, mainly involving the kidneys, but all ended in failure with the patients dying soon after the operation, or at best surviving for a few months. Doctors were aware that failure was due to an adverse reaction of the immune system, they had no way of preventing this from happening.

Determining the causes of rejection

In 1952 a French medical team led by Professor Jean Hamburger attempted the first organ transplant using a live donor – a woman whose 16-year-old son was about to lose his only kidney as a result of an accident donated one of her own. Despite the close relationship, the boy's body rejected the new organ and he died three weeks later.

On the other side of the Atlantic, Boston surgeon Joseph Murray was convinced that the key to success lay in transplanting organs between identical twins. In 1954, the opportunity arose to test his hypothesis: 23-year-old Ronald Herrick consented to donate a kidney to his brother Richard, who was dying of renal failure. Murray performed the operation with his colleagues John Merrill and Leland Harrison Hartwell. This time, the transplant was a success and Richard lived for another eight years with his brother's kidney. It was a breakthrough, but the situation was unusual and kidney transplants remained rare.

Weakening the immune system

The real challenge was less to do with clinical procedure and the technical challenge of the

Medical first *In 1960 Joseph Murray's team performed the first successful organ transplant with non-identical twins. The recipient (on the left) was given a kidney from his brother (right), and after a course of total body X-rays went on to lead a normal life.*

JEAN DAUSSET AND COMPATIBILITY

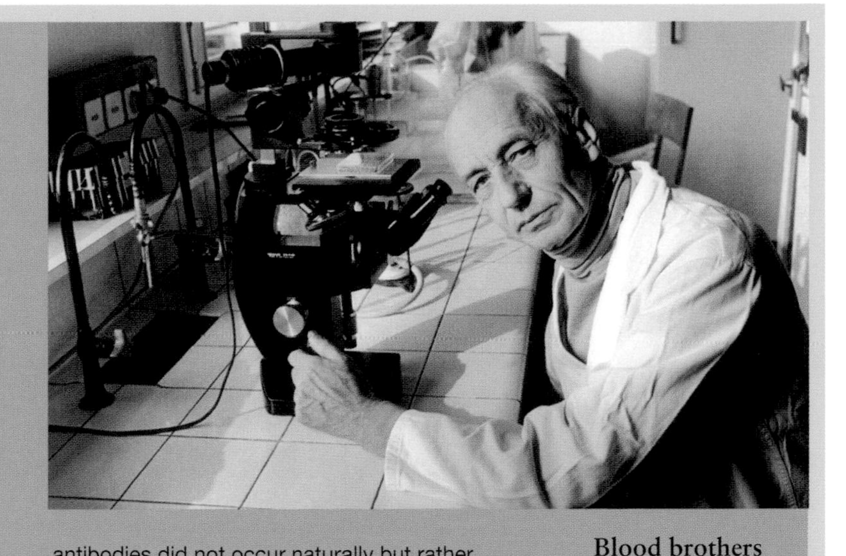

In 1952 Dr Jean Dausset (right), a specialist in blood disorders, embarked on research that would overturn the whole concept of human identity. Dausset had ascertained that some patients suffered from anaemia because their bodies manufactured antibodies that attacked their own red blood cells. He then began looking for a comparable phenomenon in white blood cells (leukocytes). Using blood serum from a woman who had received multiple transfusions, Dausset observed massive agglutinations of white blood cells.

The reaction indicated that there were different groups of leukocytes, in the same way that there are different blood groups. Yet one key difference was that, unlike the ABO blood-group system, the serum antibodies did not occur naturally but rather developed after a transfusion or a pregnancy. In 1958, by performing a series of transfusions on a patient using blood from the same donor, Dausset succeeded in identifying the first human leukocyte group, which became known as the HLA, or Human Leukocyte Antigen. Later discoveries by Dausset led to the classification of antigens into two groups, I and II. His work on histocompatibility won him the Nobel prize for medicine in 1980.

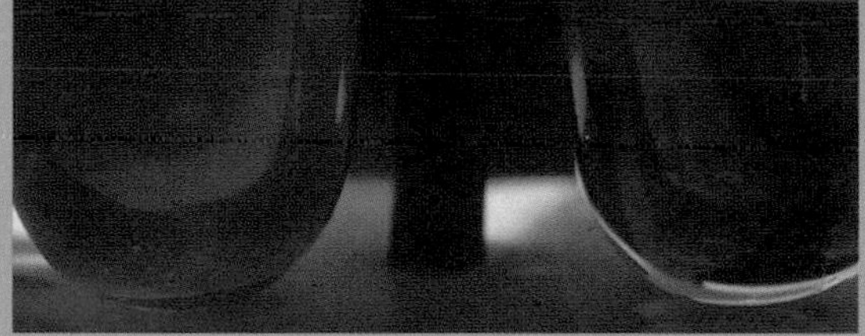

The genetic 'identity card' that the HLA system represented (in its encoding of molecules that bear the biological identity of cells and tissues of each individual) represented a major breakthrough in the understanding of the basic mechanism behind the human immune system. In effect, it was what enabled the immune system to distinguish between 'self' and 'non-self' where transplanted organs were concerned.

Thus, the closer the antigen characteristics of the donor to those of the recipient, the greater the chance of a transplant organ being accepted. In the case of imperfect compatibility, it was essential to weaken the immune system of the recipient in order to induce it not to reject the transplanted organ.

Blood brothers
Blood tests (left) are essential for gauging tissue compatibility between donor and recipient. In bone marrow transplants, for example, a patient has a one-in-four chance of being compatible with a brother or sister; with an unrelated donor, the chance of compatibility is one in a million.

operation than with basic physiology: doctors did not know how to resolve the problem of incompatibility between recipient and donor. In the late 1950s some teams began treating patients with radiotherapy and drugs known as corticosteroids to weaken the immune system, thereby lessening the chances of rejection. Gradually, more powerful immunosuppressants were developed and administered, which helped to improve the survival rate of transplant recipients. But these treatments had to be ongoing for the rest of the patient's life.

A major step forward was achieved by French immunologist Jean Dausset, whose work on histocompatibility led to more sophisticated matching of donors to recipient patients. Once the major problems of rejection had been overcome, kidney transplant techniques continued to develop in the 1960s and 1970s, and surgeons began to turn their attention to the transplanting of other organs.

The first pancreas transplant was carried out in 1966 by the American surgeons Richard Lillehei and William Kelly. The following year,

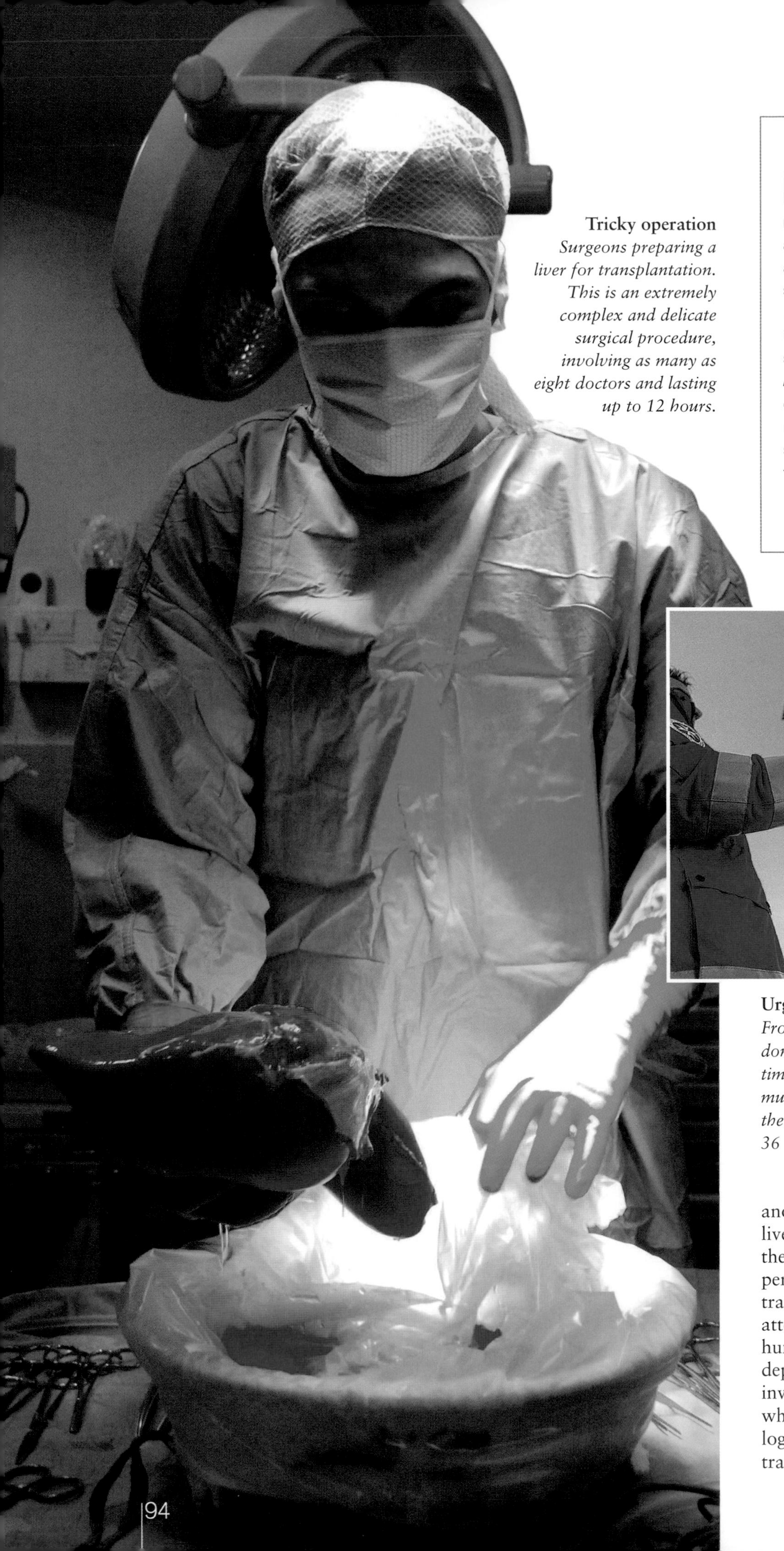

Tricky operation
Surgeons preparing a liver for transplantation. This is an extremely complex and delicate surgical procedure, involving as many as eight doctors and lasting up to 12 hours.

MAGIC MUSHROOM

In the 1960s pharmaceutical laboratories devoted a great deal of research to finding new antibiotics. Whenever scientists working on antibiotics from natural substances travelled abroad, they were in the habit of collecting samples to bring back and test. So it was that Hans Peter Frey, a biologist employed by Sandoz in Basle, Switzerland, returned in 1969 with soil samples containing the fungus *Tolypocladium inflatum*. This was found to contain a substance called cyclosporine, which turned out to be a poor antibiotic but a powerful immuno-suppressant. It was successfully tested on transplant recipients in 1978, greatly reducing the mortality rate, and is now widely prescribed as a postoperative drug for transplantees.

Urgent delivery
From the moment an organ is removed from the donor and placed in a coolbox for transportation, time is of the essence. No more than four hours must elapse before a heart is transplanted into the recipient, eight hours for a liver or lung and 36 hours for a kidney.

another American, Thomas Starzl, pioneered liver transplantation and, most famously, the South African surgeon Christiaan Barnard performed the first successful human heart transplant. One by one, transplants were attempted of almost all the organs in the human body, with more or less success depending on the technical difficulties involved. Organ transplantation became a whole new field of medicine, with its own logistical and legal framework. Excepting transplants from live donors (notably the

COMPOSITE TISSUE TRANSPLANTATION

In 1998, the French surgeon Jean-Michel Dubernard performed the world's first successful transplantation of a hand, on the New Zealander Clint Hallam. This kind of operation involves grafting many different kinds of tissue – skin, blood vessels, muscles, nerves and bones. Another medical first for Dubernard, in 2006, was a partial facial transplant on Frenchwoman Isabelle Dinoire, who had been badly mauled by her own dog. Because of the technical complexity and high incidence of rejection, composite transplants such as this are extremely rare.

Helping hand *Professor Dubernard (above right) and patient Clint Hallam after the hand transplant. Three years later, claiming he felt 'mentally detached' from his grafted hand, Hallam had it amputated.*

New face *Isabelle Dinoire before and after reconstruction of her face. By November 2006, the grafted area (nose, chin and lips) had regained sensitivity and mobility.*

Public-spirited *People willing to donate organs after their death are encouraged to carry a donor card (above). In 2008, 2 per cent of UK donors were under 16, 30 per cent aged 16–45, 34 per cent 46–60, and 33 per cent 60 or over.*

liver and kidneys), a vital aspect of organ transplantation was to define the criteria of brain death in legal terms in order to enable organs to be gathered in the best possible condition without contravening ethical guidelines.

Alternatives to transplants

Organ transplants are commonplace in the early 21st century. In the UK alone, in the year 2009–10, more than 3,700 organ transplants were carried out. Transplants have helped prolong the life of countless patients otherwise condemned to die prematurely of cardiac, renal or pulmonary disorders. Even so, despite major national campaigns to encourage organ donation, thousands of people on the waiting list die every year for want of a suitable donor. To circumvent this problem, researchers in many countries have begun looking to other solutions, such as organ transplants from animals (xenotransplantation), artificial organs or organs grown from stem cells.

SHARING BONE MARROW

Bone marrow, the soft tissue found in the interior of bones, contains stem cells that can replicate all kinds of blood cells (that is, red and white corpuscles and platelets). In certain illnesses, such as leukaemia and congenital immune deficiency syndromes, bone marrow transplants are vital operations. Mostly, a donor is sought from within the family. But failing this, doctors turn to a national bone marrow registry; over 60 countries maintain such databases and have reciprocal sharing arrangements. Marrow is taken from the donor's pelvic region under general anaesthetic and administered to the recipient in the form of a simple injection into the bloodstream. The first bone marrow transplants were conducted in 1959 by the French cancer specialist George Mathé, on a group of Yugoslavian nuclear physicists who had been accidentally irradiated.

CHRISTIAAN BARNARD – 1922 TO 2001

'Doctor of hearts'

A groundbreaking South African surgeon captured the imagination of the world when, on 3 December, 1967, he became the first person to perform a human heart transplant. Already renowned in his own country, Barnard was just as much at home in front of the cameras as he was with scalpel in hand and instantly became an international star.

'Amazing … it works!' Christiaan Barnard was overjoyed. After a nine-hour operation, assisted by a medical team of no fewer than 30 people, including his brother Marius, the 45-year-old surgeon had pulled off a remarkable feat – the world's first heart transplant. At Groote Schuur Hospital in Cape Town, South Africa, Barnard had taken the heart of a young woman, Denise Darvall, who had died in a road accident, and transplanted it into a 54-year-old grocer, Louis Washkansky, who was dying of incurable heart disease. During the operation, the patient was kept alive by a pump circulating blood around his body, while the donor heart was kept beating in a heart-lung machine. Yet Barnard's success and his patient were shortlived. Dosed with large quantities of immunosuppressant medication, which weakens the immune system to lessen the chance of organ rejection, Washkansky succumbed to a bout of double pneumonia just 18 days after the operation.

Doctor and patient *Christiaan Barnard (top) explains the medical procedure by which he transplanted the heart of a young woman into his patient Louis Washkansky (above).*

Man of the moment *By giving hope to thousands of heart patients worldwide, Barnard became a medical superstar overnight. Here he is in 1968 (left) signing autographs for a throng of admiring fans in the German city of Frankfurt-am-Main.*

Yet no surgical procedure had ever excited such interest from the press. It involved extraordinary medical skill, but what really grabbed people's attention was the fact that it was a heart, a highly symbolic organ, that had been transplanted. And the surgeon's photogenic appearance and suave, man-of-the-world attitude, stoked the media frenzy.

Global craze

A few weeks later, Barnard performed a transplant on a second patient who lived for 18 months. Others tried to emulate the South African surgeon's success: in 1968, more than 100 heart transplants were attempted in 48 countries. Sadly, most of these patients died after a few months, either through lack of knowledge on the part of the medical teams or as a result of complications. Many doctors gave up in disappointment, but despite mixed results, Barnard stuck to his guns. As with other organs, the real stumbling block was the problem of rejection. In the 1980s, the discovery of the immunosuppressant properties of cyclosporine gave new impetus to cardiac transplant operations, doubling the proportion of patients who survived for a year or more from 40 to 80 per cent. In this same period, improvement in cryopreservation methods enabled organs to be extracted at some distance from the operation site and transported to where they were needed. Throughout his life, Barnard continued to explore new techniques and was one of the first people to attempt animal-to-human transplants, a procedure known as xenotransplantation.

Barnard's eventful personal life also kept him in the news. His widely publicised love affairs earned him the nickname 'doctor of hearts'. Afflicted by rheumatoid arthritis, he was forced to give up surgery in 1983. He died in 2001 of a severe asthma attack.

PIONEER OR OPPORTUNIST?

Barnard was accused by some in his profession of 'stealing' the idea and the opportunity of performing the first heart transplant. They cited an American cardiac surgeon, Norman Shumway, as the true father of the procedure. Since the late 1950s, Shumway had conducted numerous heart transplants on dogs, but hesitated to try the operation on people for ethical reasons. At the time, all organs for transplants came from corpses. But for a heart transplant the organ had to be removed still beating, so it was necessary to extract the organ from a donor who was deemed dead. Because the concept of brain death had not been legally defined, such an operation ran the risk of outraging public opinion. On 20 November, 1967, Shumway announced that he was ready to attempt a human heart transplant. The very next day, a South African newspaper revealed that Barnard's team was also preparing. Barnard had visited Shumway in the United States and was thoroughly acquainted with his research. In the event, Shumway performed his first transplant a month after Barnard. His patient survived for a fortnight.

FORTY YEARS OF HEART TRANSPLANTS

Around 4,000 heart transplants are carried out every year around the world. The operation itself and the clinical aftercare, which by now are tried and tested procedures, have enabled some patients to survive for several decades. Even so, many cardiac patients die prematurely for want of an available organ.

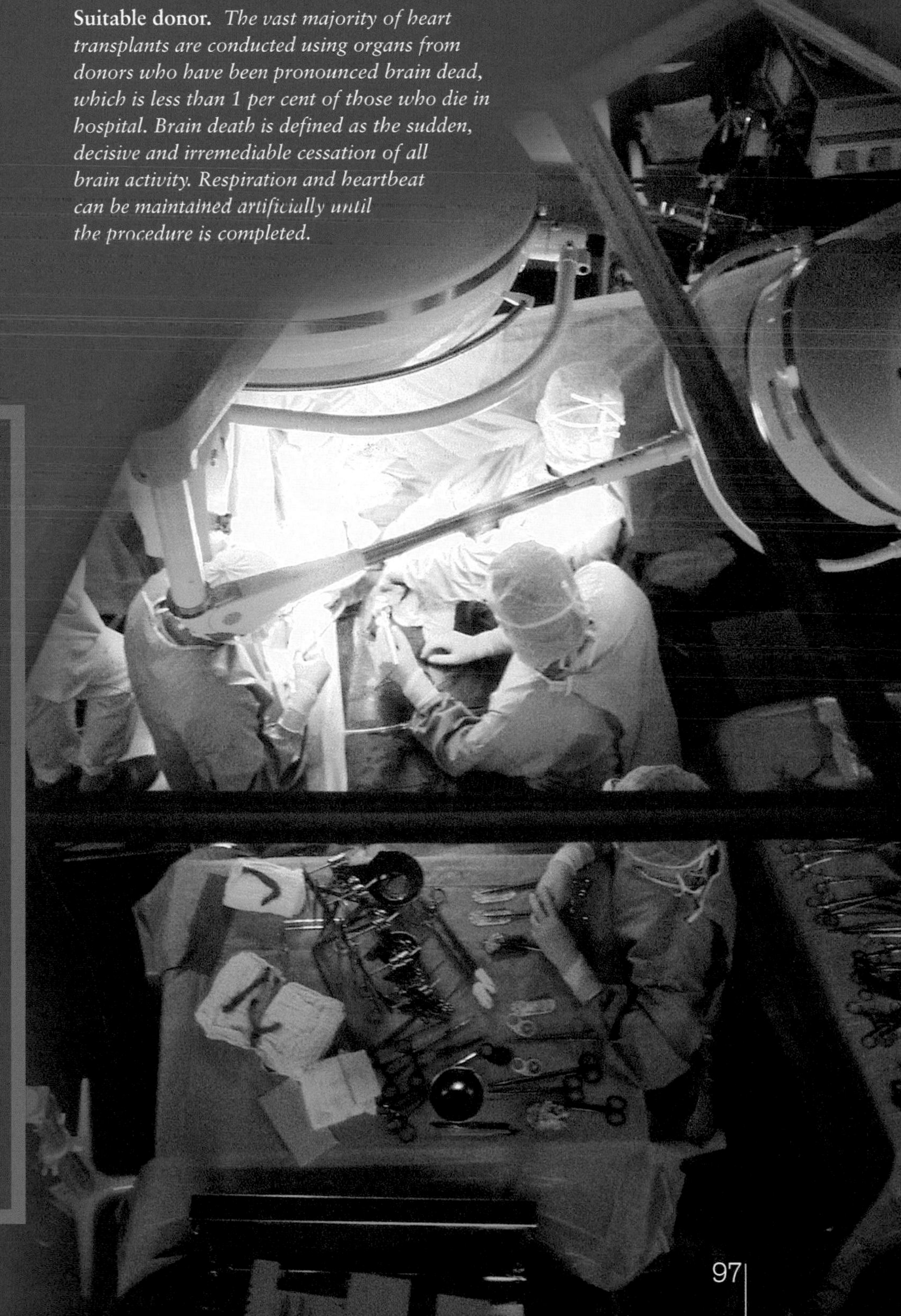

Suitable donor. *The vast majority of heart transplants are conducted using organs from donors who have been pronounced brain dead, which is less than 1 per cent of those who die in hospital. Brain death is defined as the sudden, decisive and irremediable cessation of all brain activity. Respiration and heartbeat can be maintained artificially until the procedure is completed.*

NUCLEAR SUBMARINES – 1954

Run silent, run deep

The age of the nuclear submarine began with the launch of USS *Nautilus* in 1954. Powered by on-board reactors, these vessels could stay submerged for months on end and changed the face of warfare and military strategy. Those that were equipped to carry intercontinental ballistic missiles became vital assets in the policy of nuclear deterrence and 'Mutually Assured Destruction' that defined the Cold War.

During the Second World War, the standard procedure for submarines when attacking a surface vessel was to approach as close as possible on the surface, using the submarine's diesel engines, then submerge and use electric propulsion for the final run towards the target at a low speed of 2–3 knots (3–5km/h) to conserve battery power. Once the attack was over, the sub would make a faster getaway, but even then its speed rarely exceeded 7–10 knots and this could only be sustained for one or two hours at most. In 1954 the USS *Nautilus* radically changed this game plan.

Its secret lay in its mode of propulsion: the *Nautilus* was fitted with a small atomic reactor whose uranium fuel rods turned water into steam, which in turn powered electric turbines,

PRESSURISED WATER OR LIQUID METAL?

Nuclear reactors on board a submarine work on broadly the same lines as those in power stations: as they emit neutrons, uranium atoms release heat. The main difference is in the cooling systems that prevent the fission chain reaction from running out of control. In PWRs, which are now the norm in all nuclear navies, radiation from the fuel heats a primary circuit filled with water that is kept under pressure to prevent it boiling. This heat is transmitted, via a heat exchanger, to a secondary circuit that generates non-radioactive steam. In liquid-metal fast marine reactors, which are far less common, the coolant used is liquid sodium or potassium. These give a superior performance and use less uranium, so the reactor is more compact, but they are more prone to hazardous leaks.

Polar warrior
USS Nautilus *arriving in New York Harbor in 1958 after becoming the first vessel to sail beneath the Arctic ice cap, passing over the North Pole.*

Fearsome deterrent
The cutaway drawing shows inside a Resolution*-class ballistic missile submarine from 1966, one of the first generation of Royal Navy nuclear-powered and nuclear-armed boats, carrying the US-designed Polaris missile. In the 1990s, these submarines were replaced by* Vanguard*-class boats carrying Trident missiles.*

doing away with the need to rely on battery power. As early as 1939 the US Navy had set up a working group to investigate the feasibility of nuclear propulsion, the great advantage of which lay in the fact that no air was needed to burn the fuel. The project was moribund during the war but resurfaced in 1947 under the direction of Commander Hyman G Rickover. Development of the reactor was entrusted to the Westinghouse Electric Corporation, while the keel of the vessel was laid by General Dynamics' Electric Boat Division at Groton, Connecticut. The largest wartime submarines measured just over 75 metres in length, but to carry the weight of the pressurised water reactor, naval architects had to design a hull almost 100 metres long.

Three-star comfort

The *Nautilus* was launched on 21 January, 1954. Sea trials, which began the following year, proved highly satisfactory. She could make 23 knots while submerged, and easily detected mock 'enemy' subs sent to intercept her. Her chances of survival were reckoned to be a hundred times greater than those of a conventional submarine. The *Nautilus* could stay submerged for weeks at a time, and was only forced to resurface to replenish the crew's oxygen supply.

Comfort on board was unparalleled for the boat's complement of 13 officers and 92 men. In diesel submarines the air was always fetid and humid and stank of exhaust fumes. The surplus electric power generated by the atomic reactor enabled the vessel to be completely air-conditioned and powered a desalination plant to provide abundant fresh water. With three decks, the boat had ample space for every man to have his own individual bunk.

Fitness first
Physical exercise is vital for submariners, whose tours of duty underwater can last up to three months at a stretch. Here, the captain of the USS Alabama *jogs around the missile compartment of his vessel during a period of deployment in the 1980s.*

AROUND THE WORLD IN 60 DAYS

In 1960 USS Triton made the first fully submerged circumnavigation of the globe, covering a distance of 60,000km in just 60 days and 21 hours.

Even so, the *Nautilus* was only a prototype. One abiding problem was noise. Her hull and superstructure vibrated, rendering sonar detection of other vessels ineffective when she was doing more than 4 knots.

Adding a crucial few knots

A nuclear submarine's speed derives both from its reactor power and from the hydrodynamics of the hull. In her original configuration, the *Nautilus* still had a distinct bow, which was redundant on a vessel that no longer needed to make headway on the surface. It also impaired performance when diving. The solution came in the radical design of USS *Albacore*, built in

Marking time
To combat the disorientation that can occur on long tours of duty, different colours of artificial light are used in the crew quarters to distinguish day from night: white light is for day, while red is for night time.

Dangerous encounter
On the night of 3-4 February, 2009, Britain's HMS Vanguard *(below) and the French nuclear submarine* Triomphant *– each equipped with 16 ballistic missiles – collided in the Atlantic Ocean off western France. The passive sonar these vessels use during operational patrols meant that they were unable to detect each other. This concern with secrecy came close to causing a disastrous incident that could have resulted in widespread contamination.*

1953. This conventional sub had a new 'teardrop' shape: a perfectly hemispherical bow section tapered smoothly to end in a cone at the stern. Subs with what came to be called an 'Albacore' hull proved fast and manoeuvrable.

To evade detection and pursuit, nuclear subs can dive deeper than any other large manned craft. On patrol, they generally cruise at depths of 300–500 metres, though the advanced Soviet *Alpha*-class hunter-killer subs could dive to 850 metres and reach 43 knots. To withstand the huge pressure at such depths, the Russians made these vessels broader in the beam than usual; to save on weight, from 1969 onwards the hulls were made from titanium, which is lighter than steel.

The need for silence

Noise is a crucial factor as submarines can be detected by enemy sonar picking up their sound signature. Because active sonar (which emits UHF waves) has the drawback of giving away a sub's position, passive sonar became the norm from the late 1950s on. Yet the 48 hydrophones on the USS *Nautilus*' early BQR-4 passive sonar array were drowned out by engine noise. The problem was overcome in later subs by mounting the engines and all other noise-emitting equipment on suspension systems. Pumps circulating pressurised water for the reactors have been made quieter, while the

French underwater might
The French Navy has four ballistic missile submarines in service: the Triomphant, *the* Téméraire, *the* Vigiliant, *pictured above during a presidential visit, and the latest addition, the* Terrible, *launched in 2010.*

AN ALARMING SOURCE OF POLLUTION

The operational life of a nuclear submarine is 30 years or less. Best practice dictates that after decommissioning, the nuclear reactor should be taken out of the vessel and the fuel rods reprocessed. Yet the intense arms race during the Cold War prompted the Soviet authorities to design and build nuclear subs (a total of 247) in great haste with no thought for how to deal with them once they became obsolete. Until 1991 spent reactors in old subs were simply sunk at sea. Nowadays, in ports around the Kola Peninsula in northern Russia, decommissioned Soviet subs are mothballed, awaiting proper recycling. As of 2004, there were 71 reactors awaiting reprocessing.

profile of propeller blades has been altered to minimise the noise caused by cavitation (the formation of gas bubbles).

Ballistic missile submarines

Armaments systems have also evolved. The development of ICBMs saw the emergence of a new class of subs that could act as launch platforms. Carried in vertical silos amidships, the missiles are ejected from the firing tube by a powerful jet of compressed air; the missile's own engine starts after breaking the surface. The multiple warheads can be delivered to a wide array of pre-programmed targets. The first ballistic missile submarine ('boomer' in US naval slang) was the USS *George Washington*, which entered service in 1959 and carried 16 Polaris missiles with a range of 2,200km. The Soviet Union responded with the *Yankee*-class ballistic submarines, first deployed in 1967, with a range of 2,400km and equipped with 16 SS-N-6 missiles per sub. Aside from the two former Cold War adversaries, only three other countries have ballistic nuclear subs in their arsenal: Britain, France and the People's Republic of China.

EAVESDROPPING ON THE ENEMY

One of the most audacious and valuable espionage missions of the Cold War was carried out by the US Navy's submarines *Halibut*, *Seawolf* and *Parche*. In an operation codenamed 'Ivy Bells', for a decade from 1971 onwards, these specially equipped vessels tapped into an underwater communications cable that the Russians had laid 2km beneath the Sea of Okhotsk, linking the Soviet Pacific Fleet naval base at Vladivostok with Petropavlovsk on the Kamchatka Peninsula. So sure were the Soviets that they could not be eavesdropped that they sent all their messages 'in clear' (unencrypted), providing the Americans with vital intelligence on their submarine and missile technology. The intellience coup eventually came to light when a traitor within the US National Security Agency betrayed the operation to the Kremlin.

CRUDE BEGINNINGS

The US Navy's first venture into guided missiles were Loon buzz bombs – exact copies of the German V-1 'doodlebug' – launched from the decks of diesel-electric subs from 1946 onwards. These were replaced by an American-designed missile, the Regulus I, in 1953. The Soviets experimented with a similar weapons system, the P5. But what these missiles all had in common was that the sub had to surface to launch them, which took several minutes and exposed the sub to aerial attack. Strategic naval missile deployment came of age with the advent of ICBMs that could be fired from a submerged vessel.

Test firing
Below: An Ohio-class submarine armed with Trident missiles. Below left: A Trident II missile breaks the surface. This type of ICBM, has multiple warheads and a maximum speed of 19,000km/h.

An Ohio-class ballistic missile submarine (above), launched in 1981:
- **Displacement: 18,750 tonnes**
- **Length: 170 metres**
- **Maximum beam: 13m**
- **Submerged speed: 25 knots (46km/h)**
- **Crew: 148 crew and 15 officers**
- **Armaments: Mark 48 torpedoes; either 24 Trident I or II ballistic missiles (range up to 7,360km) or 154 Tomahawk cruise missiles**

FIBRE OPTICS – 1955

Transmitting data with light

Initially used for medical applications, these bundles of glass strands, thin as a human hair, would ultimately revolutionise telecommunications. Combined with laser technology, by the late 20th century fibre optics had facilitated the rise of the information superhighway and ultrafast broadband networks

Light show
Dr Narinder Kapany (right) demonstrates his new invention in 1960 .

In 1870 the British physicist John Tyndall set up an experiment that would confound the laws of science. His aim was to prove to the Royal Institution in London that light does not necessarily travel in a straight line. His apparatus consisted of a light source, in front of which was a container filled with water. On the side facing the light, he made a small hole in the container, through which the water began to pour. As it did so, the spectators were astonished to see that the rays of light appeared to follow the stream of falling water. What Tyndall had just demonstrated with his so-called 'light fountain' was the principle of total internal reflection of light, which is the foundation of modern fibre optic technology.

It was not until 1955 that a practical application for the principle was discovered. Narinder Kapany, a researcher at Imperial College, London, was extruding filaments of glass into fine strands when he applied a light source to one end. He found that the light travelled right down the fibres to their farthest extremities, even when the fibres were bent or rolled up. It was Kapany who coined the term 'fibre optics'.The following year, gastroenterologist Basil Hirschowitz and his team at the University of Michigan lodged a patent for a flexible endoscope made from two bundles of glass fibres. After various refinements, this device helped to improve medical diagnostics immeasurably. Fibre-optic endoscopes are also used for investigating mechanical structures that are otherwise inaccessible to inspection, such as aero engines.

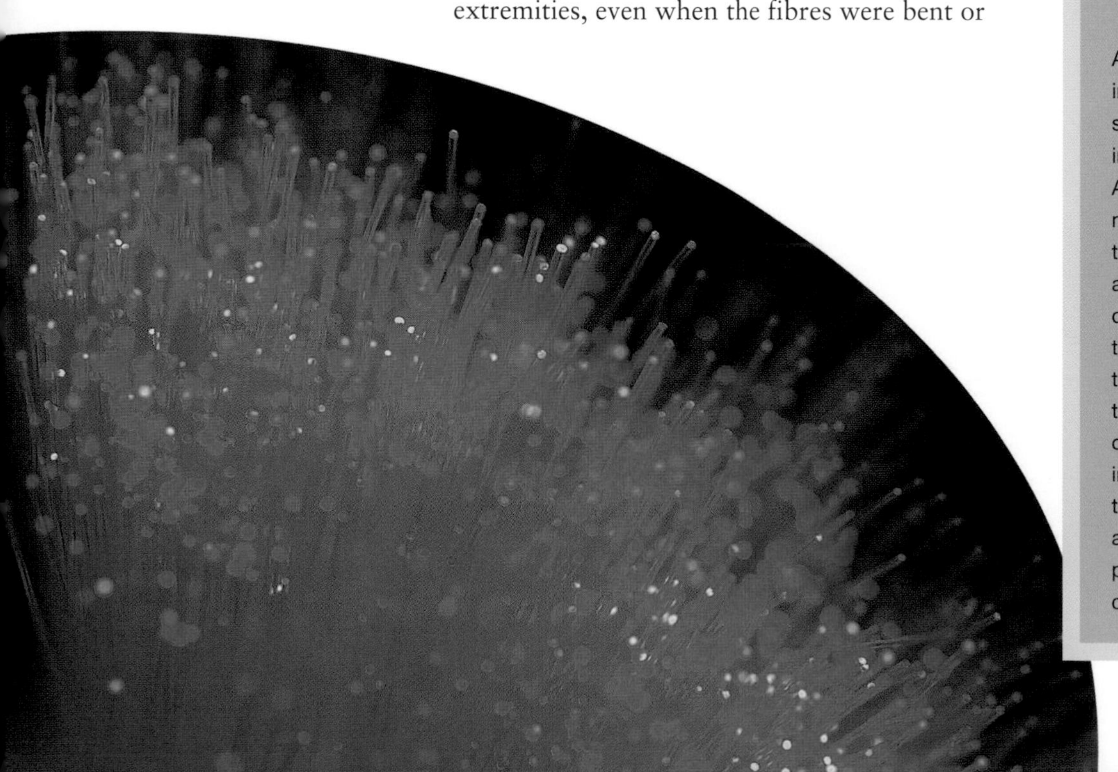

Ultrafast service
A bundle of optical fibres (below), just one of which is capable of carrying 14,000 gigabits of data every second, or the equivalent of 3,000 DVDs.

BELL'S PHOTOPHONE

The world's first cordless phone was the photophone, patented by the Scottish-American inventor Alexander Graham Bell in 1880. The device worked by transmitting sound on a beam of light. The user spoke into a horn, which set a thin mirror vibrating. Alternating between concave and convex, the mirror also dispersed and focused sunlight; these rays were detected some 200m away by a parabolic receiver, with crystalline selenium cells at its focal point. The varying resistance that the cells registered in the light signal were then converted back into sound to reproduce the message captured at the receiver. Bell considered the photophone 'the greatest invention I have ever made, greater than the telephone', but the vagaries of the weather and lack of reliable sunlight consigned the photophone to being merely a technological curiosity ahead of its time.

Past and future
Fibre optic technology has revolutionised telephone and Internet communications. On the left is an old-style cable containing 2,700 pairs of copper wires; beside it are modern optical fibre cables with 1,260 filaments.

The information superhighway

In the 1960s the advent of the laser opened up the world of telecommunications for fibre optics. This powerful source of concentrated light was capable of propagating light down much longer fibre optics than hitherto possible. The data was encoded in the form of impulses and the message was sent using digital binary language. Yet at this early stage the quality of the glass was not good enough to enable messages to be sent over very long distances, while lasers also remained unreliable.

In 1970 the Dow-Corning Glassworks of New York manufactured glass fibres of such purity that they could carry 65,000 times more data than had previously been possible with traditional copper wire and electricity. Optical fibre was not only impervious to electromagnetic surges, it was also 200 times lighter than copper cable. This development signalled the birth of the information superhighway.

En route to fast broadband

In 1977 the General Telephone Company unveiled the first telephone link using fibre optics, connecting two towns in California across a distance of 10km. That same year, the American companies AT&T, Western Electric and various Bell subsidiaries trialled the use of fibre optic cables in the states of Illinois and Georgia, as well as in Britain. Their networks became fully operational in 1978, carrying 40,000 telephone connections simultaneously in both directions. In 1985, Bell Labs achieved 300,000 simultaneous telephone conversations over a single optical fibre. Advances in laser technology continued to boost these already impressive figures over the ensuing years.

By 2008, experimental transmissions were achieving transfer speeds of 14 terabits per second. More than 25 million kilometres of optical fibres now criss-cross the globe, accounting for some 80 per cent of all telecommunications. The digital information they carry provides users with telephone links, Internet access and HD television signals. Many companies such as BSkyB and Virgin now offer combined media packages to subscribers.

Undersea link
A technician on board a cable-laying ship off the coast of Chile plays out a fibre-optic cable designed to link the country with its neighbour Peru via the Pacific seabed. This is just one of thousands of such underwater cables worldwide, vital arteries of the modern global telecommunications network.

Discovering undersea secrets

Jacques Cousteau (1910–97) devoted his life to understanding the oceans. Voyaging around the world aboard his research vessel, *Calypso*, this intrepid explorer made more than 50 expeditions to reveal the wonders and mysteries of the silent world beneath the waves. In five full-length feature films, a hundred television documentaries and more than 60 books, he passed on the wonders he discovered to the public around the world.

Master mariner
Commander Jacques Cousteau (below) during his service with the French Navy. In later years, on his expeditions, he took to wearing a red woollen hat, which became his trademark. In 1992, he was appointed a consultant on the environment for the United Nations and World Bank.

Jacques Cousteau was born on 11 June, 1910, in the town of Saint-André-de-Cubzac near Bordeaux in southwest France. Despite suffering from enteritis and anaemia, in 1930 he gained a place at the prestigious naval academy at Brest. Calling at far-flung ports while serving aboard the training vessel *Jeanne-d'Arc*, Cousteau took every opportunity to use his ciné camera, a hobby he had been keen on since the age of 13. In 1933 he graduated as the second-best cadet in his year. But though he was drawn to the sea, the young naval ensign felt an even greater urge to become a pilot, and so enrolled at the Naval Aviation College at Hourtin in the Gironde. While he was there, a serious car accident forced him to give up flying.

Filming the depths

To help regain strength in his badly injured arm, Cousteau took to swimming in the rocky inlets east of Marseilles. On these outings he took with him a 9.5-mm camera, ingeniously mounted inside a glass preserving jar to keep out the seawater, which he used to take underwater photographs. When war came, he served with distinction in the Resistance, in contrast to his elder brother Pierre-Antoine who collaborated with the Vichy régime.

In 1943 Jacques and the engineer Émile Gagnan devised the aqualung, an underwater breathing device consisting of a steel cylinder of compressed air and a pressure-control valve

Alien beauty
The coral reefs in the Red Sea off Egypt were the subject of the Calypso*'s first research voyage in 1951.*

FIRST OF MANY

Cousteau made his first underwater film in 1942, a 17-minute documentary entitled *Par dix-huit mètres de fond* ('18 Metres Deep'). It was made off Embiez, in the south of France, and involved Cousteau diving without breathing apparatus and filming with the aid of a depth-proof camera case specially devised by the engineer Léon Vèche, who had been a naval cadet in Brest with Cousteau. The film won first prize at the Congress of Documentary Film in 1943.

(regulator) that supplied air at normal pressure to the diver. It allowed the wearer to stay submerged for much longer periods than using a simple snorkel. It also rendered obsolete the traditional heavy diving suit and helmet, in which the diver was supplied with air via a line from a ship or shore-mounted pump. Cousteau's partnership in this venture earned him a 5 per cent commission on all sales. As the Aqua Lung company grew to become the world's leading supplier of scuba diving equipment, he amassed a small fortune, which he used to fund both his own research and other scientific and technological projects worldwide. It also allowed him to maintain a top-class team of experts. In 1946 he was one of the co-founders of the Underwater Study and Research Group (GERS) based in the naval port of Toulon in the south of France. Three years later, he helped to salvage Professor Jacques Piccard's bathyscaphe *FNRS-2* when it foundered off Dakar in West Africa.

Cousteau's *Calypso*
The former British minesweeper which Jacques Cousteau used for his expeditions was accidentally rammed and sunk by a barge in Singapore harbour in 1996. Quickly salvaged, she was taken back to France for a total refit and relaunch as a museum ship, but is currently languishing in a shipyard amid legal disputes over unpaid repair costs.

Cousteau and the *Calypso*

In 1937 Cousteau had married Simone Melchior, the daughter and grand-daughter of French

admirals. The couple had two sons, Jean-Michel (born 1938) and Philippe (1940), who both grew up to accompany their father on his expeditions. With his business thriving, in 1949 Cousteau set out to find a suitable ship to convert into a research vessel to test out the new diving and underwater photography technology he had developed. He found it in the *Calypso*, a former Royal Navy minesweeper that had been converted into a ferry on Malta, then more recently acquired by the wealthy brewer Sir Thomas Loel Guinness. Guinness agreed to lease the *Calypso* – named after a sea-nymph in Homer's *Odyssey* who held Odysseus captive on the Maltese island of Gozo – for a symbolic 1 franc per year.

It took two years for the naval shipyard at Antibes in the south of France to refit the 43-metre wooden-hulled *Calypso* as an oceanographic research vessel. The latest equipment was installed, including a gyrocompass controlled by an autopilot, a highly accurate echosounding device, state-of-the-art radios, and an underwater viewing chamber beneath the ship's bows, with five portholes for filming.

Cousteau placed the *Calypso* at the service of the French Navy and it remained the country's only oceanographic survey vessel up until 1965. Simone, who accompanied Cousteau on his voyages, prided herself on being 'the only sailor's wife who waits on board ship for her husband'. The crew nicknamed her 'the Shepherdess'.

THE SILENT WORLD

In 1956 Jacques Cousteau and the French filmmaker Louis Malle made *The Silent World*, a documentary based on the bestselling book of the same name, which Cousteau had published in New York three years earlier. It was the first documentary film to use colour underwater footage, shot in the Mediterranean, the Red Sea, the Indian Ocean and the Persian Gulf. It became the first documentary film to win the Palme d'Or at the Cannes Film Festival; it also won the Oscar for the Best Foreign Language Movie of the year. The stars ofthis full-length feature film, which ran for 86 minutes, included 'evil' sharks, 'good-guy' sperm whales, comic seals and a grouper called 'Merou'. The critics hailed it as a triumphant updating of the world of Jules Verne. Interviewed by journalists at Cannes, Cousteau confidently predicted that 'in just a few years, man will journey to the deepest depths of the ocean'.

Curious craft
Cousteau's Denise *(bottom left) was the prototype for all subsequent exploration submersibles and ROVs. This 'diving saucer' was fitted with a floodlight to illuminate the sea bed and a manipulator arm for picking up objects.*

THE DIVING SAUCER

Cousteau was always at the cutting edge where underwater filming was concerned. In 1959 he and the engineer Jean Mollard constructed the *SP-350*, a small saucer-shaped submersible, which was given the nickname *Denise*. It was capable of diving to a depth of almost 400m and could carry two crew, lying side-by-side on mattresses to view the ocean floor through forward-facing portholes. *Denise* measured 2.85m across the beam, weighed 3.5 tonnes and was equipped with an ingenious propulsion system. Seawater was drawn in and then squirted out under pressure through two movable, electrically powered jets, enabling the craft to move like a squid at a speed of around 2 knots (*c.* 3.7km/h).

Sea station off Marseilles
Nicknamed 'Diogenes' after the Stoic philosopher who lived in a barrel in ancient Greece, this steel vessel was both home and laboratory for the first oceanauts in Cousteau's 'Conshelf' experiments.

AT THE HELM OF A MUSEUM

In 1957, at the instigation of Prince Rainier, Cousteau was made director of the Monaco Oceanographic Institute, a research centre founded in 1910 by Rainier's father Albert I, a noted amateur oceanographer. Cousteau was given the task of publicising the results of his expeditions and making them available to scholars worldwide. He stayed at the helm of the institute for 31 years. Its large aquaria are home to more than 350 species of fishes (from 6,000 specimens in total), 200 species of invertebrate and 100 different tropical corals.

Marine habitats
The Monaco Oceanographic Institute has meticulously re-created several different marine environments in its tanks. This picture, (above) shows a section of coral reef habitat containing sharks, rays and a variety of smaller tropical fishes.

The adventure begins

The *Calypso* set sail from the naval yard at Toulon on her maiden voyage of exploration on 24 November, 1951. On board were a scientific team plus a film crew that included the director Jacques Ertaud. Their destination was the Red Sea, to study the coral reefs there. Cousteau's first expedition was funded in part by the National Geographic Society. It was his firm conviction that, in order to really understand the ocean and to enthuse the general public about the magical undersea world, it was essential to study and film its wonders at first hand. Three years later, he returned to the same location, financed this time by an oil prospecting contract for the Gulf of Oman. Further expeditions took him and his team to the far corners of the world: the Antilles, Madagascar, Polynesia, New Zealand, Cuba, the St Lawrence River, China's Yellow River, Antarctica and the Amazon. Each voyage provided marine biology with invaluable new insights.

Cousteau resigned his naval commission in 1957, with the rank of corvette captain. His growing reputation as an environmentalist was boosted by his condemnation of the dumping of barrels of radioactive nuclear waste in the Mediterranean in 1960. The same period saw Cousteau instigate various undersea dwelling projects. In his 'Conshelf' experiments in 1962, for example, two 'oceanauts' stayed for a week in a small underwater steel 'house', just 5 metres long and 2.5 metres across, anchored at a depth of 10 metres off Marseilles. The following year, the 'Precontinent II' experiment in the Red Sea involved two compartments, one 9.5 metres below the surface and the other at a depth of 25 metres. In 1965, in his most ambitious experiment in underwater living, Cousteau assembled a 6-metre diameter steel sphere, 'Precontinent III', 110 metres below the waves off Cap Ferrat. On board, six men who

Leading from the front *Cousteau and members of his team preparing to dive. They are using the aqualung he designed, which gave divers greater freedom of movement and allowed them to remain submerged for far longer periods.*

had trained like astronauts lived for six weeks, breathing a mixture of helium and oxygen that made their voices sound like Donald Duck.

Cousteau's big TV break

In 1966 the US television network ABC commissioned Cousteau to make a series of 12 documentary films, each lasting an hour. It paid the princely sum of 400,000 dollars a programme. As well as tapping into his huge passion for filming and commentary, this lucrative contract gave Cousteau the funds to organise further expeditions to investigate all the areas he was interested in – zoology, geology, geophysics, marine biology, archaeology and ecology. The first programme, which was broadcast in 1968, was about his exploration of Lake Titicaca. The lean, emaciated figure of Captain Cousteau, with his hooked nose, omnipresent red hat and pronounced French accent, soon became a familiar sight to TV audiences in America, then around the world. He presented his programmes not as dry documentaries but as real-life adventures, captivating adults and children alike. His firm belief was that knowledge had to excite people before it could educate them. His popularity grew steadily with each transmission. The scientific community was also extremely open-minded towards this self-made man, recognising in him a gifted communicator who was making science accessible to ordinary people.

By the late 1960s, Cousteau was one of the most famous people in the world. He devoted most of his energies to producing and distributing his films, and made it his main mission to introduce viewers worldwide to the glories of the sea. The resources he controlled afforded many scientists the possibility of exploring hitherto uncharted regions of the world's oceans.

Deep-sea camera *One of the cine-cameras mounted in special underwater housings that Cousteau and his team developed in order to film their groundbreaking documentaries.*

FAILED VENTURES

Not all of Cousteau's projects were a success. In the late 1960s he embarked on an ambitious scheme to build a submarine to explore the deepest ocean trenches: a 300-tonne vessel with a double-skinned hull, which he named the *Argyronete*. The project collapsed in 1972, costing the French taxpayer 57 million francs. And the Cousteau Oceanic Park, headed by his son Jean-Michel, went bust shortly after opening in 1989 in the former market at Les Halles in Paris.

Elegant hunter
A shark swimming with dolphins over a coral reef in the Indian Ocean. Cousteau was one of the first people to present a positive image of sharks as beautiful animals. Up till then, these threatened species had been demonised en masse as ruthless man-eaters.

'Captain Planet'

In the 1970s and 80s, Cousteau began to focus on the deterioration of the Earth's marine environment, which he considered a vital common reserve for ensuring humanity's survival when terrestrial resources had been exhausted. In 1990 he was instrumental in bringing about an international agreement that placed a 50-year moratorium on all mineral exploration in the Antarctic. In 1992, at the UN Earth Summit in Rio, his sponsorship of a petition, signed by 9 million people, to protect the rights of future generations earned him the affectionate nickname 'Captain Planet'.

Cousteau died on 25 June, 1997, at the age of 87. He left behind him two organisations, The Cousteau Society and the *Équipe Cousteau*, to which he bequeathed exclusive rights to use his name, image and archives, including all of his films and photographs. Both are dedicated to raising funds to help to protect the environment.

THE CAPTAIN'S TRADEMARK

Jacques Cousteau's iconic red woollen hat was supposedly modelled on the watch cap worn by convicts in the former penal colony at Toulon. These men were often employed as 'heavy-suit' divers in the port, and wore the caps to protect their heads inside the cumbersome brass diving helmets.

THE HOVERCRAFT – 1955

Travelling on a cushion of air

To commemorate the 50th anniversary of the first flight across the Channel by Louis Blériot, the *SR-N1* left Calais for Dover on 25 July, 1959. Neither a plane nor a boat, this craft certainly flew – but only a few centimetres above the waves. On board, its inventor Christopher Cockerell savoured his triumph. After all, his invention very nearly failed to get off the drawing board.

Working model *Christopher Cockerell built this small prototype hovercraft in 1955 to demonstrate the feasibility of air-cushion vehicles (above right).*

Christopher Cockerell had tested the principle of the hovercraft in 1954, using a makeshift apparatus consisting of two food tins of different diameters, an industrial air blower and a pair of kitchen scales. Placing the small tin inside the larger one, he attached them to the nozzle of the blower before turning the whole affair upside-down and clamping it to a stand, so that it hung down resting lightly on the flat weighing surface of the scales. When he switched the blower on, the air stream from the gap between the tins created an air cushion that raised the smaller tin slightly, decreasing the weight registering on the scales. The benefits of such a system were immediately apparent. A craft using the principle would be freed from drag created by the friction between the hull and the water, thus gaining a huge advantage in speed.

The idea was not a new one; as far back as 1877, English inventor John Thornycroft had patented the concept of a vessel with a concave-shaped base to its hull, which could be filled with air to reduce hydrodynamic drag. And in 1929 the massive German Dornier *Do-X* flying boat put Thornycroft's theory into practice, when, during an Atlantic crossing, it flew much closer to the ocean's surface than was usual in order to take advantage of the air-cushion effect. The aircraft's performance was boosted as a result and the journey time significantly reduced.

Cool reception

Cockerell patented his invention in 1955 and set about canvassing support for the hovercraft from aircraft manufacturers and shipyards alike. But he found no takers; the former saw it as a boat, while the latter thought it was an aeroplane. Undeterred, the inventor went ahead and built a small prototype, founding his own company, Hovercraft Development Ltd, in 1956. The only response of the British military establishment when he tried to interest them in the hovercraft was to place the vehicle on the 'classified' list to prevent the idea from falling into foreign hands. Eventually, in 1958, the Ministry of Supply awarded the frustrated Cockerell a contract to build a full-scale prototype.

Taking off

The aircraft and seaplane manufacturer Saunders-Roe Ltd of Cowes on the Isle of Wight was commissioned to build Cockerell's first hovercraft, the *SR-N1*, a small machine just 9 metres long and 7 metres wide. Air was drawn in by a vertically mounted piston engine and blown out beneath the craft, where it formed a cushion. Later, to contain the air

Sea skimmer *The world's first practical hovercraft, the* SR-N1, *during trials. In 1961 a flexible skirt was added to the craft's rigid hull.*

VERSATILE VEHICLE

The hovercraft's all-terrain capability has made it synonymous with adventure. In fiction, hovercraft have appeared many times in James Bond movies, beginning with *Diamonds Are Forever* (1971) and continuing through *The Spy Who Loved Me* (1977), *Moonraker* (1979) and *Die Another Day* (2002). In real life, French explorer Michel Peissel used one to cross the Himalayas in 1973. In 2000 hovercraft were even used as icebreakers in northern Canada. Their ability to skim over water and ice floes makes them ideal for transporting people and freight in inhospitable northern latitudes such as Alaska and the Arctic Circle.

cushion on even ground, the *SR-N1* was fitted with a skirt made from a tough rubberised fabric. This also helped to raise the vehicle further off the surface of the water. After being fitted with an auxiliary aft-facing jet engine, the *SR-N1* achieved a speed of 50 knots, almost 100km/h, in 1961.

The first regular hovercraft service across the Channel was inaugurated between Ramsgate and Calais by the Hoverlloyd Company in 1966. Two years later, a rival company, Seaspeed, began operating from Dover. By 1980, when the service was at its peak, hovercraft were carrying 2.4 million passengers and 389,000 vehicles annually, accounting for 31 per cent of all cross-Channel traffic. The 40-metre long *SR-N4*s could each carry 254 passengers and 30 cars; over time, this capacity was increased (by adding a new centre section) to 418 passengers and 60 cars. On average, the crossing took 35 minutes, compared to 90 minutes for the shortest ferry crossing. Hovercrafts continued crossing the Channel until 2000, when escalating fuel costs and competition from the ferries and the Channel Tunnel, which had opened in 1994, brought the service to an end.

Action man
Hovercraft seemed almost tailor-made for James Bond. In the film Die Another Day, *director Lee Tamahori staged an exciting chase scene involving 007 piloting a small hovercraft.*

Swift crossing
From 1968 to 2000, SR-N4 *passenger and vehicle-carrying hovercraft operated a service across the English Channel from Dover and Ramsgate to Calais and Boulogne. Pictured here is Seaspeed's* The Princess Margaret. *Her sister vessel,* The Princess Anne, *holds the record for the fastest ever hovercraft crossing of the Channel: 22 minutes, on 14 September, 1995.*

WELL PRESERVED

The longest-running commercial hovercraft operation is the 10-minute ferry link across the Solent from Southsea to Ryde on the Isle of Wight, which has been in service since 1965. At Lee-on-the-Solent more than 100 hovercraft from past eras are preserved in the world's only hovercraft museum.

Synthetic diamonds 1955

As diamonds are both rare and expensive, over the years many inventors have tried to manufacture them artificially. In 1878 the Scottish chemist James Ballantyne Hannay claimed to have made synthetic diamonds by bringing a mixture of lithium, bone oil and paraffin to red heat in a sealed wrought-iron tube. After 82 attempts – and several explosions – he finally obtained some minute crystals; the British Museum verified them as diamond, but neither he nor anyone else could repeat the result.

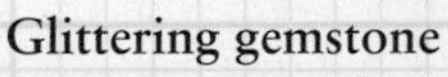

Glittering gemstone
All diamonds, whether natural or artificial, result from the same chemical process, the crystallisation of carbon, and have the same chemical and physical properties.

THE ULTIMATE MEMORIAL

The latest trend in synthetic diamonds is to make them from the cremated remains – or even hair – of the dearly departed. This burgeoning industry was started in 2001 in the USA by LifeGem®. The graphite released during cremation is transformed into coloured diamonds ranging from 0.25 to 1.3 carats. In a bizarre twist, the company is now creating diamonds from a lock of the singer Michael Jackson's hair. The hair had been in the possession of collector John Reznikoff, who acquired it 25 years earlier from Ralph Cohen, executive producer of the Jackson Pepsi Cola commercial. Cohen took the lock in January 1984, while rescuing the star from an accident that ignited his hair.

The search continues

In 1893 the French chemist Henri Moissan believed he had synthesised diamonds when he superheated charcoal powder, iron and graphite inside a carbon crucible in an electric arc-furnace, then plunged the resulting alloy into either water or molten lead. But what he had actually produced was silicon carbide (carborundum). None of these early experiments was able to generate the intense heat and pressure that is required to make diamonds.

In the 1940s Harvard physicist Percy Bridgman re-created the high pressure (over 100 tons per cm^2) and temperature (3,000°C) found deep within the Earth. Even so, he failed to produce diamonds and the project was shelved until after the Second World War.

The first reproducible synthesis was finally achieved by Swedish company ASEA in 1953, but the first commercially successful synthesis was made in the USA in December 1954 by Tracy Hall of General Electric. Hall's breakthrough came from the use of a 'belt' press producing pressures approaching 100 tonnes per cm^2 and temperatures above 2000°C. Graphite was dissolved with molten nickel, cobalt or iron, which in turn helped to speed the conversion of carbon to diamond. The largest gem Hall created was just 0.15mm across and too small for jewellery, but it gave rise to the industrial man-made diamond industry with applications as diverse as the cutting heads of drills and styluses for record turntables. Synthetic gem-quality diamond crystals were first produced in 1970.

'Diamonds are a girl's best friend'
Marilyn Monroe delivering her famous song about the permanence of diamonds and the fickleness of lovers in the 1953 feature film Gentlemen Prefer Blondes.

The atomic clock 1955

New accuracy
American physicist Charles H Townes demonstrating an early atomic clock to a student in 1955 (left). Twelve years later the International Bureau of Weights and Measures redefined the duration of the second as 9,192,631,770 cycles of radiation, corresponding to the transition between two energy levels of the caesium-133 atom.

In 1955 the British physicists Louis Essen and Jack Parry of the National Physical Laboratory in Teddington built a device that used the electronic transition frequency of atomic spectra to revolutionise timekeeping. Their method, pioneered by the American Norman Ramsay (and earning him the 1989 Nobel prize for physics), entailed combining a caesium-beam resonator with ultra-high-frequency quartz crystal oscillators to calibrate atomic time with astronomical time scales. This was the world's first atomic clock, 100,000 times more accurate than a quartz clock.

A new time standard

Meanwhile, in 1954–6, Jerrold Zacharias of the Massachusetts Institute of Technology (MIT) developed the first commercial atomic clock. He and a local company marketed the device as the 'Atomichron'. Zacharias also planned to build what he called an 'atomic fountain', a revolutionary type of atomic clock accurate enough to study the effect of gravity on time that had been predicted by Einstein. This idea was resurrected in 1989 by a team from Stanford University, who made the world's first working caesium fountain using a technique called laser cooling. This involved trapping a cloud of caesium atoms at the intersection of six laser beams and reducing its temperature to a few millionths of a degree above absolute zero. The cloud was then launched up through a cavity in which microwave radiation caused the atoms to oscillate. The atoms' pulse was taken at the top of their trajectory. The most precise atomic clocks yet created are those at the US National Institute of Standards and Technology (NIST), accurate to 1 second in a billion years. In 2004 NIST produced an atomic clock that was smaller than a grain of rice

Such precision has brought about major advances in many fields, including geophysics, geodesy, astronomy, astrophysics and cosmology. The benefits of the atomic clock are evident to the general public in global positioning system (GPS) devices such as Sat Nav units in vehicles.

200 ARE BETTER THAN 1

To ensure even greater accuracy, International Atomic Time is an average of the time kept by more than 200 atomic clocks in some 70 national laboratories worldwide.

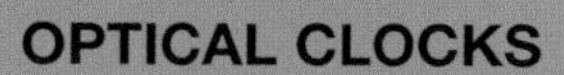

OPTICAL CLOCKS

Whereas atomic clocks use microwaves to measure atomic transitions, a new generation of so-called 'optical clocks' are being developed which use light instead and these can be more than 100 times as accurate. They work by trapping ultra-cold atoms within a lattice of intersecting laser beams and using them to calibrate the frequencies of electromagnetic waves.

Extreme miniaturisation
A chip-scale atomic clock, built by NIST in 2004, played a major part in increasing the accuracy of later GPS systems.

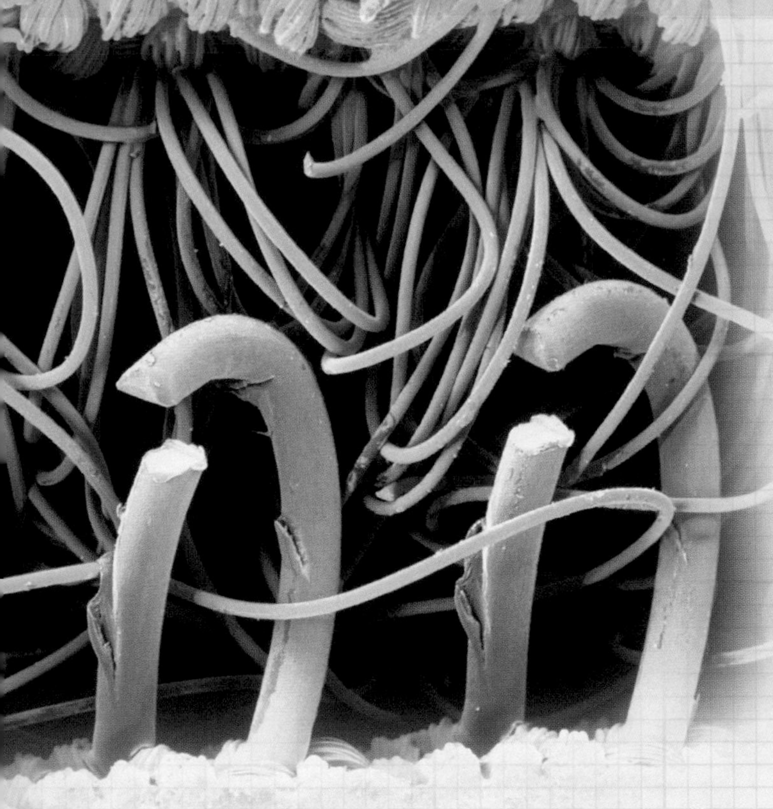

Velcro 1955

Returning from a country walk one day in 1941, Swiss engineer George de Mestral noticed that several burdock burrs had attached themselves to his clothes and to his dog's coat. Looking at them under a microscope, he saw that they had thousands of tiny hooks. This gave him the idea of creating a fastening system made from two strips of nylon fabric, one covered with hooks and the other with loops. While capable of holding a garment together firmly, they had the advantage of being easy to rip open. It took Mestral ten years to perfect his invention. He patented it in 1951 in Switzerland and in 1955 in the USA, under the name 'Velcro', an abbreviated compound of the French words *velours* and *crochet* ('hook'). Today, millions of metres of Velcro are sold every year, for use on clothing, shoes and bags.

Rivalling the zip
Velcro viewed under an electron microscope, showing the hooks are in blue and the loops in green. When the two layers are pressed together, the hooks catch the loops, thus holding the strips together. After being ripped apart, the nylon hooks and eyes quickly regain their original shape, allowing the Velcro to be used over and over again.

Lego 1955

The Lego story began at Billund in Denmark, in a joiner's workshop run by Ole Kirk Christiansen. He had been making wooden toys since 1932 and in 1934 gave his products the name 'Lego', from the Danish phrase *Leg godt*, meaning 'play well'. In 1949 the company manufactured the first of its famous construction-toy bricks. Christiansen based his design on the self-locking toy bricks called 'Kiddicraft', introduced by the British toymaker H H Fisher Page two years earlier. Lego bricks, made from cellulose acetate, harked back to traditional wooden building blocks, but unlike wooden ones could be joined together in many different ways using an ingenious interlocking system of round studs on top and hollow bases. This made them easy to assemble and to pull apart again. Since the company began selling its product in Europe in 1955, it has manufactured more than 327 million bricks. Lego is now available in 130 countries worldwide, and there are four Legoland theme parks, in Denmark, England, Germany and California.

MOVING WITH THE TIMES

To keep ahead in the competitive world of children's toys, Lego has had to constantly innovate. Since 1996 the company has opened new R&D facilities in Milan, Tokyo and London, and brought out Lego specifically aimed at girls, such as doll's houses, a Lego video game, and a CD-ROM. It has worked with the Massachusetts Institute of Technology to develop 'Toys that Think' (right), which incorporate programmable bricks that can interact with the physical world through sensors and motors.

Disposable nappies 1956

Most people associate disposable nappies with Pampers, which were developed by American chemist Victor Mills for his employer Procter and Gamble using his own children as guinea-pigs. From a first prototype produced in 1956, the product was launched in 1961 and has become the market leader. But it was a British woman, mother-of-three Valerie Gordon Hunter, who brought the world's first disposable nappy to market. She and her husband, Pat Hunter, obtained a UK patent for her invention in 1948, then teamed up with the manufacturer Robinsons of Chesterfield to produce Paddis, which were launched at the Mothercraft Exhibition in London in November 1950.

A MOUNTING PROBLEM

When they were introduced, disposable nappies were seen as a luxury to be used on days out. But as incomes have risen and convenience has become more prized, they have become the nappy of choice for almost all parents. This has led to a growing ecological problem. In the UK alone, more than 3 billion disposable nappies are thrown away every year. The plastics in disposable nappies take between 200 and 500 years to biodegrade, compared to 6 months for a traditional washable cotton one.

Big business
Its is estimated that the average baby goes through between 5,000 and 6,000 nappies from birth to becoming potty-trained.

Go-kart racing 1956

Under starter's orders
Go-karts lined up ready to race in 1959. Go-karts developed from soapbox carts, crude homemade vehicles that relied solely on gravity for speed.

The father of go-kart racing was the American Art Ingels. In 1956, while working for the American carmaker Kurtis Craft in Glendale, California, Ingels spent his free time building a small car from pieces of scrap metal and a 2.5-horsepower lawnmower engine. The car could run at almost 50km/h, but for the driver, who sat low down virtually on the tarmac, the sensation of speed was breathtaking. The idea caught on, and by 1958 purpose-built karts were rolling off the production line of the newly-formed Go Kart Manufacturing Company. These were powered by chainsaw or outboard motors. Karting was introduced to Europe in 1959, where the sport became an instant craze.

NURSERY FOR TALENT

As go-karting grew in popularity and the machines became faster, the sport became a proving-ground for budding racing-car drivers. In the 1970s Nelson Piquet and Ayrton Senna (right) and more recently Lewis Hamilton, all of whom went on to become Formula 1 World Champions, cut their teeth on go-kart racing. Modern superkarts can reach top speeds of over 260km/h.

THE CONTRACEPTIVE PILL – 1956

Effective birth control

In liberating women from unwanted pregnancies, the contraceptive pill gave them control over their bodies and their choice of lifestyle. Its impact went far beyond the scientific and medical domain. Indeed, the advent of the Pill heralded nothing short of a social revolution.

Preventing pregnancy
The first contraceptive pills to go on the market in the 1960s (above) had an oestrogen dosage of between 50 and 150 micrograms, as compared with just 20 to 35 micrograms in modern equivalents. These lower-dosage pills are effective at blocking ovulation while also reducing potential side effects, such as blood clots (venous thrombosis).

Since time immemorial, it was common wisdom that a woman who had already conceived could not get pregnant again. Yet this seemingly banal fact was only explained scientifically in the late 19th century, when the Scottish embryologist John Beard and the French physician Auguste Prenant showed that women do not ovulate during the term of their pregnancy.

The growth in physiological research in the first decades of the 20th century gave scientists further insight into this phenomenon: the hormones released during pregnancy, oestrogen and progesterone, actively block ovulation. A welter of studies conducted at this time revealed the mechanisms that controlled and regulated the female reproductive cycle.

A slow gestation

Foremost among these studies were those by the Austrian physiologist Ludwig Haberlandt. In 1921, in tests using rabbits, he succeeded in temporarily sterilising individual females by transplanting into them ovaries extracted from other, pregnant animals. This led him to conceive of the idea of a hormonal contraceptive, and to set about developing a product that he labelled Infecundin. Eventually, after much difficulty, he managed to get a pharmaceutical company interested in it, but no doctor could be persuaded to run trials of the drug on women. In the interwar years, any birth-control procedure attracted the opprobrium of the medical establishment, and in most jurisdictions was against the law.

The chemistry of hormones

For the time being, then, contraception remained largely beyond the pale of permissible research subjects for the pharmaceuticals industry. But it was believed that many other applications could derive from work on sex hormones – in particular, remedies for menstrual disorders and for certain types of infertility.

These hormones were quickly pinpointed: the American chemists William Allen and

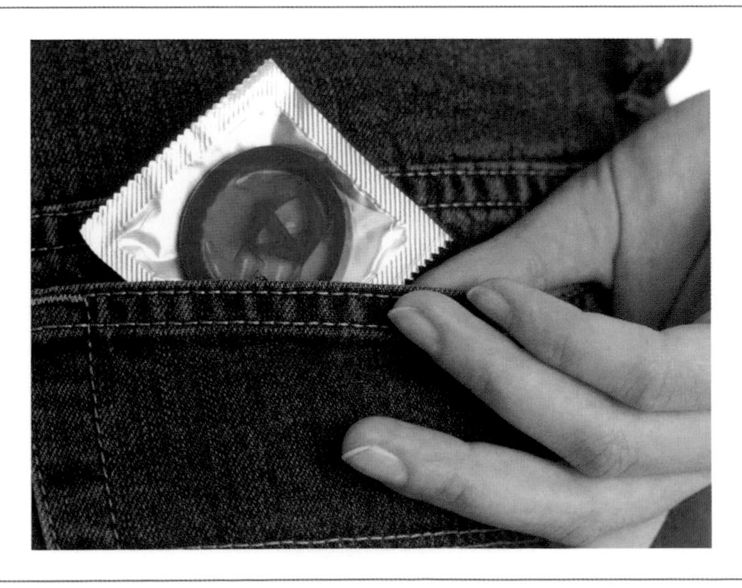

BEFORE THE PILL

In the 19th century, the most common methods of contraception were *coitus interruptus*, sponges impregnated with alcohol, vinegar or oil, and vaginal douches after the event. Over time, more elaborate methods were devised. These included the latex prophylactic (1850), which had a bad image, being associated with the prevention of venereal diseases; spermicides in the form of jellies, powders or pessaries; the rubber diaphragm, introduced in 1881 by the German gynaecologist Wilhelm Mensinga in 1881; and the interuterine device (IUD) known as the coil, the invention of another German, Ernst Gräfenberg, in 1928. There was also the so-called rhythm method, developed by Japanese obstetrician Ogino Kyusaku in 1929, which involved learning to recognise the fertile period of the menstrual cycle and abstaining from sex for its duration. But all of these methods were restrictive and unreliable, with a high failure rate. Sometimes, the only recourse left to women was to have an illegal 'backstreet' abortion, a highly risky procedure that often ended in tragedy.

Edward Doisy were the first to identify oestrogen in 1923; Allen and George Corner discovered progesterone six years later. This sparked a race between companies to devise an industrial process for extracting large quantities of these substances. In 1934 Roussel Laboratories patented a method for isolating oestrogen from the urine of pregnant mares, while in 1941 the American chemist Russell Marker achieved the first practical synthesis of the pregnancy hormone progesterone by using extracts from a wild yam species that grew near Veracruz in Mexico.

A feminist issue

At this time, the American birth-control campaigner and feminist Margaret Sanger, who knew that scientists now had the wherewithal to create a cheap and easy-to-use contraceptive pill, began lobbying in favour of its introduction. In 1950, she met the biologist Gregory G Pincus, a renowned expert in the field of reproductive physiology and head of the Worcester Foundation for Experimental Biology. With over 40 years' experience in promoting the cause of women's rights, Sanger had little trouble in persuading Pincus to embark on hormonal contraceptive research.

Invaluable support for this project came from the suffragist, philanthropist and trained biologist Katherine McCormick, who in 1952 put part of her considerable personal fortune at Pincus' disposal.

The ideal contraceptive

The task facing Pincus' research team was very clear: now that the antiovulatory properties of progesterone were known, it was down to them to discover a synthetic form of this hormone that would be effective in small doses and could be taken orally.

After testing hundreds of synthetic hormones on animals, in 1954 the team finally lighted upon two that displayed all the right characteristics – namely, norethindrone (produced by the Syntex Laboratories) and norethynodrel (synthesised at Serle Labs). It was this latter hormone that became the active ingredient in the first pill tested on women. The first clinical trials of the new pill, on a group of 12 women, took place in Boston, Massachusetts in 1956, under the supervision of gynaecologist John Rock. More extensive studies were then conducted in Puerto Rico. These tests showed that Pincus' pill was effective in almost 100 per cent of cases, and that its effects were completely reversible: the women's fertility was in no way impaired long-term, while children conceived after they

A PIONEERING CAMPAIGNER

In 1912 Margaret Sanger was working as a nurse and midwife in the working-class districts of New York City. One day, while visiting a poor woman who was desperate not to go into labour for a fourth time, she heard the doctor say: 'If you want to avoid another pregnancy, tell your husband to go sleep on the roof!' A few weeks later, the woman was dead after a botched self-induced abortion. Appalled at the inflexible attitude of the political and religious establishment to unwanted pregnancies, Sanger began campaigning for free family-planning advice for women, an illegal act in the USA at the time. Her belief that birth control was vital to women's health and to happy marital relationships led her to publish pamphlets, give lectures and open the first clinic dispensing advice and contraceptives to couples in 1916. She was prosecuted on several occasions, and used the furore surrounding her trials to publicise her ideas. For her, the advent of the Pill was the crowning achievement of a life entirely devoted to the struggle for the rights of women.

Fighting the cause
Margaret Sanger (1879–1966), pictured here (below) in 1916, at the time of her trial for opening a family planning centre. Her counterpart in Britain was Marie Stopes, a Scottish campaigner who opened the UK's first family planning clinic in London in 1921.

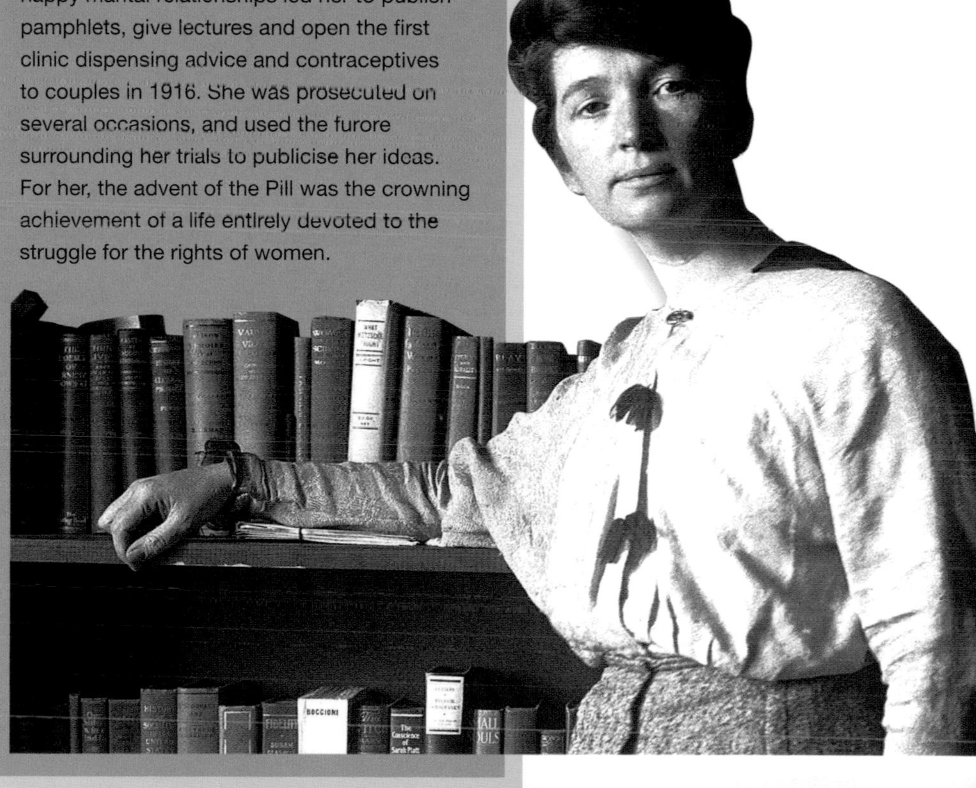

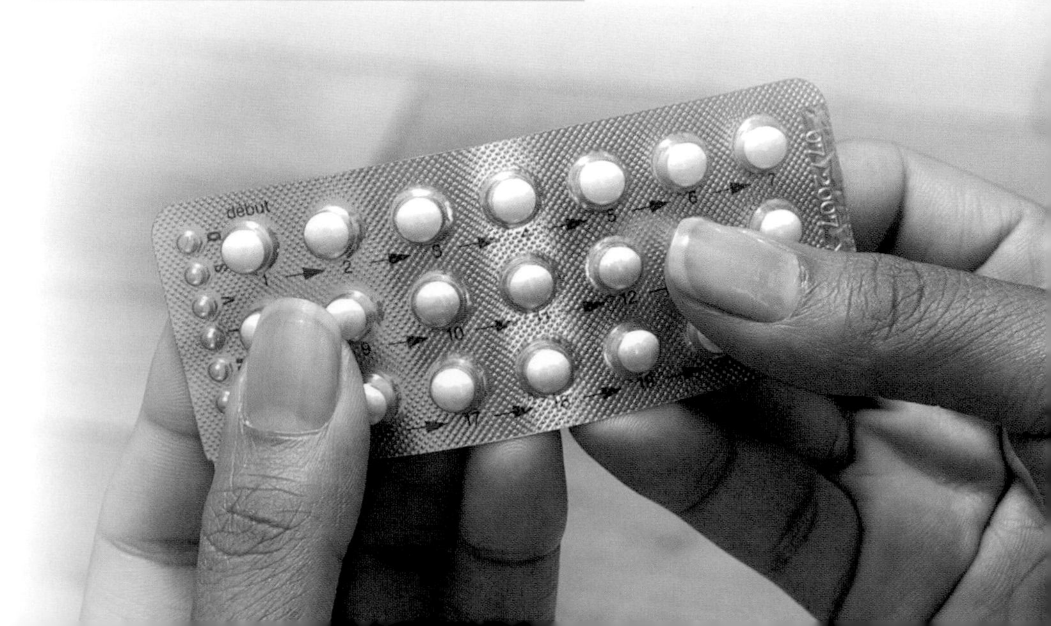

Contentious subject
An Italian poster (above) proclaims 'Enough! All it takes is one little pill'. In countries such as Italy and Spain, where the traditional teachings of the Catholic Church held sway, the women's movement encountered fierce resistance to its demand for free contraception to prevent unwanted pregnancies.

A woman's right to choose
Protesters during the 1960s and 70s in Britain (above) and France (right) demonstrating in favour of contraception and abortion on demand. This period saw women becoming politically active in demanding control over their own bodies and the right to enjoy sex for pleasure, not just procreation.

stopped taking the Pill displayed absolutely no abnormalities. The few minor side effects that did manifest themselves, such as headaches and slight nausea, were overcome in time by altering the dosage in the Pill.

A bitter pill to swallow

The contraceptive pill was authorised for general sale in the United States in 1960. It was an instant success, and its usage steadily soared. It went on the market the following year in Australia, West Germany and Britain, and it was legalised in France in 1965 and Canada in 1969.

Even so, its introduction was decried by those who feared that it would open the floodgates to a general decline in morals. More recently the use of condoms, with or without the Pill, has been advocated in a drive to reduce the risk of AIDS. Accordingly, in some countries, legalisation of the Pill was hedged around with numerous restrictions. In the USA, for example, it could only be prescribed for 2 years. Fierce debates ensued in the 1970s between conservatives and liberals before the final legal and ethical barriers were overcome and women were granted the right to have the Pill on demand.

The Pill provoked a root-and-branch change in moral attitudes. In sexual matters, people became far more open and relaxed about their chosen lifestyle. But it also had a major impact on the labour market. Since women could now choose for themselves when, or even whether, to have children they were free to gain an education and pursue careers on an equal footing with men. The Pill changed society beyond recognition, giving women control of their bodies, their relationships and their professional lives.

CONTRACEPTION WORLDWIDE

Taking the contraceptive pill is now part of everyday life for more than 100 million women worldwide. It is the most common form of contraception in Europe. In global terms, the most popular methods of birth control are sterilisation (39 per cent), the IUD (23 per cent), followed by the pill (12 per cent) and condoms (8 per cent). In developing countries, where access to family planning may be difficult, long-lasting contraceptives such as injections and coils are favoured where they are available.

Dispensing advice
A Kenyan doctor advises a woman on taking the Pill (above). According to the World Health Organisation (WHO), 137 million couples around the world use no contraception, despite the fact that they want greater control over the timing of pregnancies. WHO and national governments have therefore put in place programmes of family planning advice and free access to contraceptives.

COMPUTER-ASSISTED MANUFACTURING – 1956

A new form of automation

Once controlled by a human operator, machine tools were found to produce a more precise and reliable result when computers were put in charge. In response to an increasing demand for diversity, computer-assisted manufacture (CAM for short) enabled factories to move from high-volume standardised mass production of a single item to smaller series runs.

Man and machine *From the 1960s, as computer-assisted manufacture became the norm in industry, a cutter operating a milling machine became an increasingly rare sight in factories. In this new automated world, human input was confined to programming the computers that then controlled the movements of machines on the factory floor.*

MADE BY ROBOTS

The American physicist and engineer Joseph F Engelberger developed the original industrial robot, the Unimate, in the 1950s. The first to be installed on a factory floor was at General Motors in 1961, but it was the Japanese who were destined to develop Engelberger's advances to the full.

The classic scene you might encounter on walking into a factory in the early 20th century was as follows: in an atmosphere of deafening noise, an array of different machine tools would be drilling, cutting, planing and shaping components as they moved slowly down the assembly line. None of the workers fashioned individual parts by hand. Each man (or woman) would be seated or standing behind their allotted machine, pressing buttons, rotating flywheels or operating levers.

The machines kept up a fast and relentless pace, mass-producing standardised items. But in the post-war period, with the growth of the consumer society and the emphasis on individual taste, demand began to become more diversified. Whatever the particular product – from cars to domestic appliances to office equipment – manufacturers had to offer their customers a wider choice of models in different colours and equipped with different features.

Automated machine tools

The key element that enabled factories to respond to this new demand came in the form of computer-assisted, numerically controlled (NC) machines. These were the brainchild of the American manufacturer John T Parsons. During the Second World War, as a supplier to the aircraft industry, his company turned out complex parts in which absolute precision was at a premium. In 1946 he hired one of the earliest computers made by IBM to perform various engineering calculations; the idea was that the punched cards it produced would contain coordinate position data for the milling-machine operators to follow when cutting the surface contours of airfoil shapes. In practice, although the data that Parsons and his chief engineer, Frank Stulen, fed in was precise, the operators could not match their accuracy. So Parsons and Stulen hit upon the idea of using the punched cards to activate servomotors that directly controlled the milling machines. The US Air Force put up the funding for the venture in 1949.

To equip his machine with a real-time servomechanism that could constantly correct the movements of the cutting heads, Parsons approached the Massachusetts Institute of Technology (MIT). But following a dispute about funding, MIT negotiated a contract directly with the Air Force that effectively excluded the Parsons Company from further development of the device. The MIT-designed machine, unveiled in 1952, had a 7-track punched-tape operating system controlling its

COMPUTER-AIDED DESIGN

The first computer-aided design (CAD) system, the IBM 2250, was launched in 1964, but it was the 1980s before CAD was widely adopted. The American company 3D Systems developed a system that could produce plastic prototypes of objects designed by CAD. This involved polymerising liquid plastic into the computer-designed shape using a UV laser beam steered by a computer. Nowadays, CAD software is employed in many sectors, notably the car and aerospace industries and in architectural practices.

movements – these were more efficient than punched cards, which could become inverted during operation. The development was hailed in the journal *Scientific American* as a revolution in manufacturing techniques: 'With such machines in control, we can conceive of factories which will process, assemble and finish any article of manufacture.'

Changing the face of the factory

After the US Air Force gave its full backing to the development of NC machines in 1956, Douglas Ross, a computer scientist at the MIT Servomechanisms Laboratory, devised a programming language that would enable a computer to control the sequence of movements performed by machine tools. His Automatically Programmed Tool System (APT) signalled the birth of computer-assisted manufacturing (CAM). The huge advances in computer technology in the 1970s streamlined and lowered the cost of CAM. At the same time, the growth of computer-aided design allowed technicians to study and amend virtual prototypes before embarking on production. Automated manufacturing soon became the norm in factories. Machine-tools controlled by a single computer were clustered together to perform a variety of tasks in the production process, facilitating ever more complex, accurate and higher quality components. Skilled machine-tool operators became largely a thing of the past as manufacturing companies now required computer-literate technicians to control operations instead.

Digital modelling
Above: Industrial designers employed in R&D use specially dedicated CAD software packages to model components prior to production. Above left: A spark plug modelled on a 3-D system.

Computerised workplace
A technician at a Ford engine assembly plant in Britain programs a manufacturing operation.

Artificial kidneys 1956

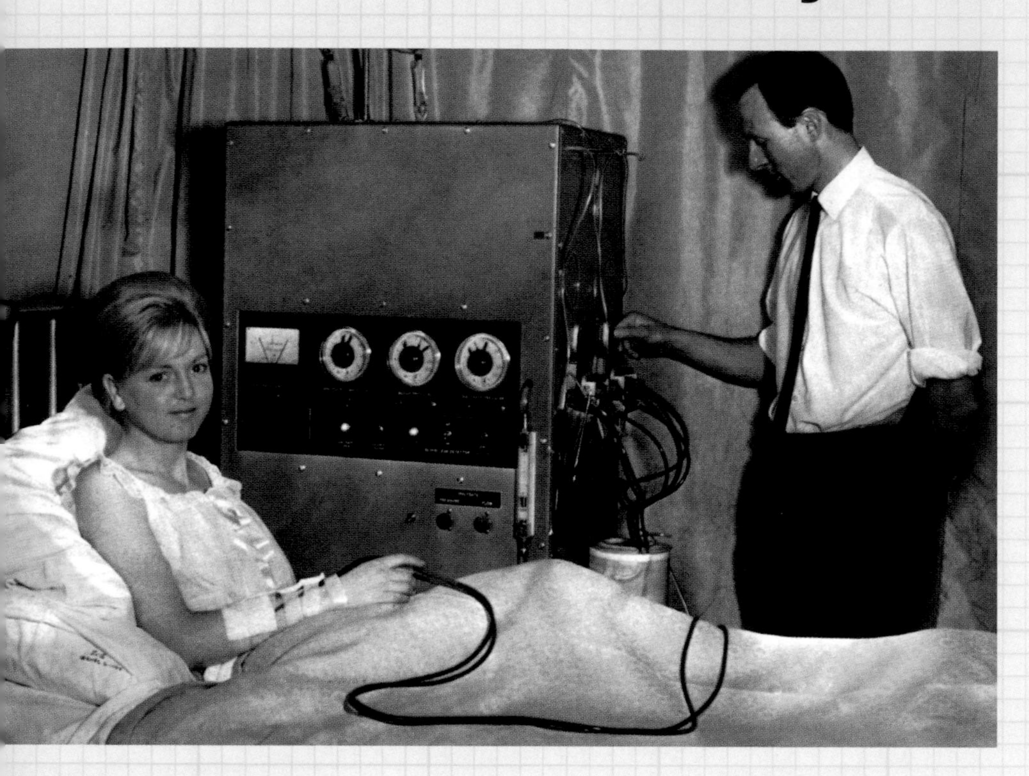

Bulky lifesaver
The first kidney dialysis machine that could be used at home was introduced in Britain in 1966. It was developed jointly by the Queen Elizabeth Hospital in Birmingham and the firm of Joseph Lucas Limited. It is seen here (above) being used by a couple from Solihull.

Up until the mid-20th century, kidney failure could be fatal. In 1938 the Dutch physician Willem Johann Kolff unearthed details of experiments in kidney dialysis conducted on dogs in 1913 by the American researcher John Abel. The experiments had used the principle that when two solutions of different concentrations are separated by a semi-permeable membrane, over time they exchange molecules through the membrane so that their concentrations become the same. Kolff applied the idea to a process to purge the blood of toxins and excess urea, using a kidney machine – comprising a semi-permeable tube dipped in a dialysing solution – which cleansed the patient's blood outside the body.

Cellophane and anticoagulants

In 1943 Kolff trialled a new type of membrane made from cellophane. This involved passing the blood through a long tube wound around a rotating drum, the lower portion of which was immersed in a tank of artificial serum. He used the anticoagulant heparin to prevent blood clots forming. After 16 failed attempts, he finally saved his first kidney patient in 1945.

Over the following years, a number of other researchers addressed the problem of kidney dialysis. In Sweden, Nils Alwall added a suction pump to the dialysis circuit. Meanwhile, Kolff had emigrated to the United States and approached Baxter Healthcare Laboratories to help him develop his device. The first commercial models appeared in 1956.

Advances in materials, miniaturisation and automation helped to establish dialysis as the principal method of treating cases of acute or chronic renal failure, when a transplant was not possible. Even so, the procedure is arduous: the patient undergoes sessions lasting between 3 and 5 hours three times a week. To free sufferers from haemodialysis, future hopes rest on implantable bioengineered kidneys grown from donated renal tissue.

Cleansing the blood
Even modern kidney dialysis machines are cumbersome affairs. In 2009 a prototype of a portable dialyser was launched in the USA, comprising a 5kg belt that filters the blood continuously. Trials are still ongoing.

ENSURING BLOOD FLOW

Since 1966, it has been standard practice to extract the blood for renal dialysis from an arteriovenous fistula. This procedure involves the surgeon connecting a vein in the patient's arm to an adjacent artery, thus causing the vein to dilate and the blood flow to increase, which makes dialysis more effective.

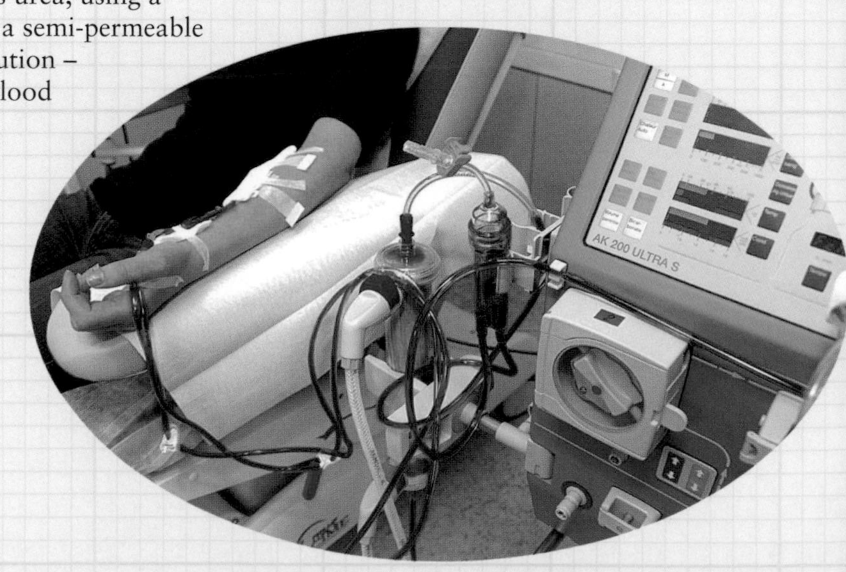

Barbie® dolls 1956

Toy innovation
Ruth Handler (below) who created the first adult-bodied doll for the children's toy market.

In the summer of 1956 the Handler family were holidaying in Switzerland with their children, Barbara and Kenneth. Back home in the USA, they owned a small toy firm, Mattel, which specialised in making accessories for dolls. One day, they noticed a number of tiny dummies in a shop window, all dressed in different fashion outfits. Ruth Handler bought one to give to her daughter. The doll in question was German-made, about 30cm tall and named Lilli. She was based on an original cartoon-strip character of the same name created by artist Reinhard Beuthien for the tabloid newspaper *Bild*.

Three years later, at the International Toy Fair in New York, Mattel launched the Barbie® doll. It was an instant hit, with her shapely figure and huge number of outfits to dress her in – including jewellery, shoes and clothes inspired by creations premiered on the catwalks of Paris. Like the originals, these outfits kept pace with changing fashions. Barbie® became much more than a children's doll, held up by some as the epitome of feminine beauty while decried by others as a stereotype of women as objects.

Fashion statement
When the original 'Ponytail Barbie'® appeared in 1959, she sported a zebra-striped swimsuit and was available as a blonde or a brunette. In 2009 Mattel revamped the brand with their 'Fashionista' range of dolls (right).

A GROWING FAMILY

Barbie® was soon joined by a family of little brothers and sisters (Skipper, Tutti and Todd), a boyfriend (Ken), plus various cousins and friends. Mattel also equipped their star product with a suitably glamorous dream home, luxury cars and other lifestyle accessories designed to enthral little girls worldwide.

Super Glue 1958

The proud claim made by the manufacturers of Super Glue was that it would rapidly stick any object, however fragile or heavy, and never come unstuck. The chemical secret behind the wonder adhesive was cyanoacrylate. The molecules of this compound only had to come into contact with water in a minute quantity – such as atmospheric moisture – to congeal and harden instantly, thus firmly bonding together the surfaces it was applied to. The adhesive properties of cyanoacrylates had been discovered by accident in 1942 by the American chemist Harry Coover. In 1951, when his colleague at Eastman Kodak, Fred Joyner, encountered the same phenomenon, they decided to develop the substance as a new glue, which came on the market as Super Glue in 1958.

Today, cyanoacrylates are also employed in medicine, for mending certain tissues, and in criminal investigations, where film of the adhesive is used to lift fingerprints off objects.

The silicon chip arrives

Miniature miracle
A printed circuit board with nine microchips. These tiny plastic or ceramic casings with pins conceal minuscule integrated circuits encased within wafers of silicon.

On 12 September, 1958, engineer Jack St Clair Kilby demonstrated a new kind of circuit he had created, comprising a single transistor and supporting components on a single slice of germanium, connected to an oscilloscope. A week later, he produced a memory cell on the same lines. This was the birth of the integrated circuit.

In 1957 Jack Kilby, an engineer working for Texas Instruments, was struck by a highly original idea: rather than making transistors one at a time, why not make several together by overlaying different impurities and placing them on the same piece of semiconductor material? His line manager was sceptical; standard practice then was to join diodes, capacitors, resistors and other components into circuits using external wires. At the time, Kilby did not realise the full importance of his idea, which was to usher in a whole new era of electronics, culminating in the PC.

Bringing together several innovations

Strictly speaking, the integrated circuit was not an invention in its own right. The novelty lay in the inspired yet intensely practical notion of accommodating several components on a single chip. Eventually, the concept would be developed to the point where several hundred million or even billions of transistors are contained on just a single microchip.

The integrated circuit represented the culmination of several discoveries, including the junction transistor introduced by Bardeen and Brattain in 1947, and also the planar process for making transistors out of silicon rather than germanium, invented by the Swiss scientist Jean Hoerni in December 1957. Even so, despite not being a physicist himself, it was Kilby who received the 2000 Nobel prize for physics for his momentous innovation, the first such occasion in the history of the award.

Nowadays, the integrated circuit, universally known as the microchip, is an indispensable part of life. They are found in all the electronic equipment so widespread

Inventor of the chip
Jack St Clair Kilby holding photographic film of an integrated circuit blueprint in 1965.

JUST BY CHANCE

As Jack Kilby himself acknowledged in his lecture marking the award of the 2000 Nobel prize for physics, good luck played a large part in his discovery of integrated circuits. By combining components (capacitors and transistors) on a single germanium substrate, he was 'just the first person to have this good idea, working with good materials and at the right time'. Kilby was a prolific inventor at Texas Instruments, with over 60 patents to his name, and he went on to invent the electronic portable calculator and the thermal printer. In a further crucial development of the integrated circuit, in 1959 Robert Noyce (who later helped found Intel) and Jean Hoerni of the Fairchild Semiconductor Corporation devised planar technology. This process involved spreading different layers onto a silicon wafer to make a flat transistor, then evaporating metal strips onto the surface to make the necessary connections.

in modern society: computers, mobile phones, chip-and-PIN bank cards, domestic appliances and sensors in car engine management systems. Many animal owners have microchips injected under the skin of their dogs, cats or horses to help retrieve their pets if they get lost or stolen.

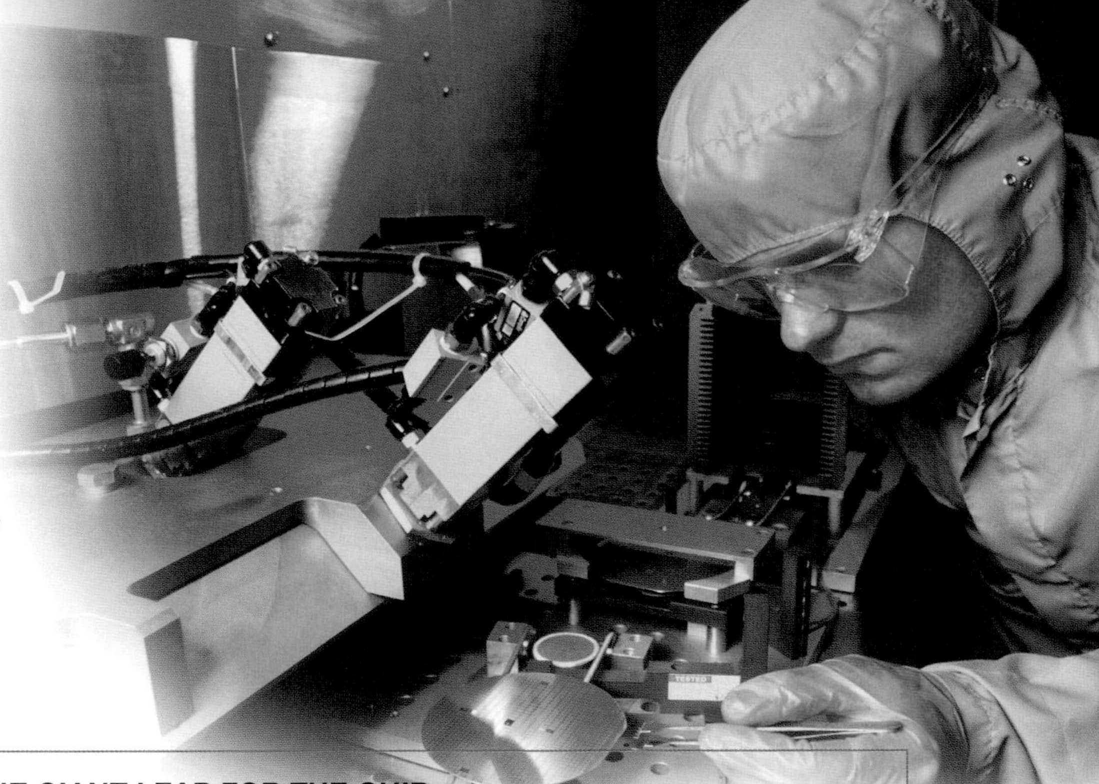

Delicate construction
A technician monitoring the fabrication of silicon-based integrated circuits. Individual chips are constructed on a single wafer of crystalline silicon (visible at bottom centre). The successive processes involved in manufacturing a chip are carried out in a temperature-controlled environment, free of all dust and humidity.

ONE GIANT LEAP FOR THE CHIP

The first steps taken by US astronaut Neil Armstrong on the Moon on 21 July, 1969, will no doubt live in most people's memories as the most momentous event in the history of space exploration. But for electronics experts, they also marked the beginning of the integrated circuit industry. Two computers called AGC (Apollo Guidance Computers) – one installed in the lunar module (LM), the other in the command module – had been built for the mission and they were the first to use integrated circuits. Each weighed 32kg, which was extremely lightweight for the time. NASA sourced the 4,000 circuits it needed from the Fairchild Semiconductor Corporation, an order which at the time represented no less than 60 per cent of the world's total integrated circuit output. The computers themselves, which were developed at the Massachusetts Institute of Technology in Boston, were designed to help the astronauts perform various operational tasks such as calculating trajectories and distances and to act as autopilots for the modules. As it turned out, during the Moon landing the AGC in the lunar module failed, forcing Neil Armstrong to take over manual control of the craft and guide it to a touch down 4 miles (6km) away from the planned landing site.

Information highways *Copper conductive pathways on a printed circuit board (PCB). These pathways connect the different layers of the circuit. At bottom right is the corner of a microchip on the board.*

The snowmobile 1959

In 1923, aged just 15, Joseph-Armand Bombardier of Valcort in Quebec had the idea of mounting the engine of his father's old Model-T Ford on a large sleigh and fitting it with a wooden propeller he had designed and made himself. The young inventor could not have known that another Canadian, Joseph-Adalbert Landry, had lodged a patent for the world's first snowmobile that very same year – a slow, cumbersome, caterpillar-tracked vehicle that stayed in production until 1948.

In 1934 Bombardier's son could not be got to hospital quickly enough in the depths of winter and subsequently died. This personal tragedy spurred Bombardier to solve the problem of getting around on snow effectively. He created a traction system for a vehicle that involved tracked rear wheels driven by a rubber-covered geared sprocket wheel.

The first Bombardier B7s (denoting the number of people each could carry) rolled off the production line in 1937. Their success peaked in a major order from the Canadian government during the war. After 1945, the firm responded to a growing demand for all-weather vehicles to take children to school, but following a succession of mild winters, the Quebec government began organising snow clearance measures instead. This prompted Bombardier to seek out new markets.

The introduction of smaller, lighter engines in the 1950s, plus the double caterpillar track designed by Joseph-Armand's son Germain, finally enabled the company to create the vehicle that its founder had long dreamed of: the Ski-Doo snow-scooter, which was launched in 1959. It was originally called the 'Ski-Dog', but a sign painter made a mistake on the first prototype and the name stuck. Equipped with wooden skis, an all-rubber track and a four-stroke engine, the Ski-Doo could reach speeds of 40km/h. Fur trappers, mineral prospectors and Canadian first-nation peoples provided its initial market. Over time, the Ski-Doo was taken up by extreme-sports enthusiasts, and sold in large numbers far beyond Quebec.

Polar dash

In 1968 an American insurance salesman, Ralph Plaisted, organised the first motorised overland expedition to the North Pole. On 7 March, 1968, Walt Pedersen, Gerry Pitzl Jean-Luc Bombardier (Joseph-Armand's nephew) and Plaisted himself set off on Ski-Doos from Eureka on Ellesmere Island. After a journey of 1,330km, they reached the North Pole on 19 April .

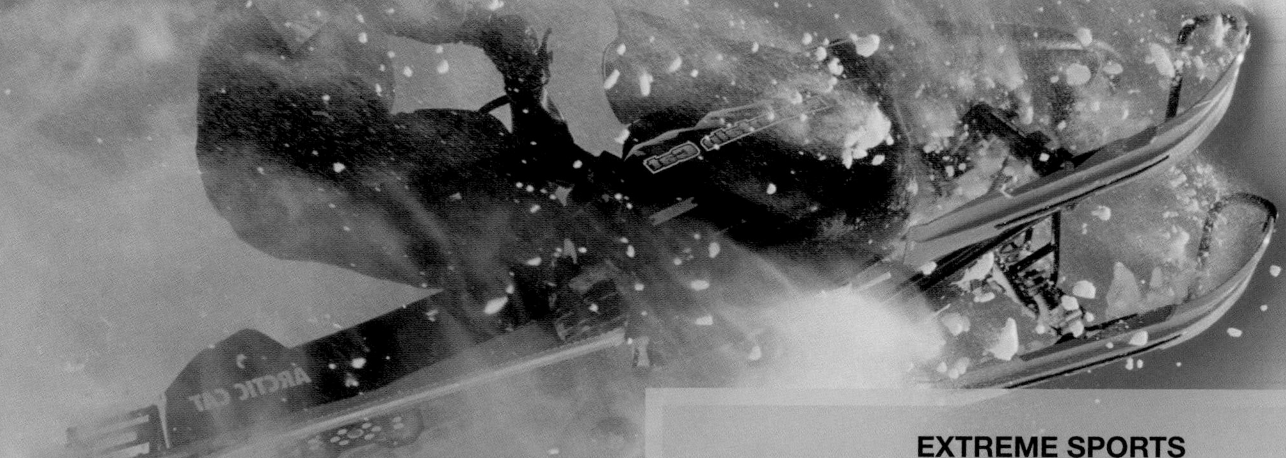

EXTREME SPORTS

Sports snowmobile riding takes various different forms, from cross-country on an oval course to downhill 'snow-cross'. It is extremely popular in North America. The Canadian Formula 1 drivers Gilles Villeneuve and his son Jacques have both been snowmobile champions and the family has continued interest in the sport.

Seatbelts 1959

In a head-on crash in a car doing 50km/h, a person weighing 75 kilograms becomes a projectile of 2.5 tonnes. These stark statistics explain why seatbelts are such an essential safety feature in motor vehicles. As early as 1903 the Canadian inventor Gustave Désiré Lebeau submitted a patent application for an automobile safety harness. The two-point safety belt (the lap type still used by passengers on airliners) made its first appearance in 1958 on a production-model car, the Saab GT750. The three-point seatbelt familiar in cars today was designed by the Swede Nils Bohlin and first fitted in the Volvo P544 in 1959. Inertia-reel belts, also introduced by Volvo, made their debut 10 years later.

Wearing a seatbelt was made compulsory in France in 1973 (outside built-up areas), in Germany in 1976 and in Britain in 1983, after a long and impassioned debate in the House of Commons. Seatbelts in the back of cars became a more common feature in the 1980s. These also became compulsory in many European countries in the ensuing years.

LIFE-SAVING ARGUMENT

Some people still refuse to wear seatbelts in cars. Anti-seatbelt campaigners often cite the accident that killed the actor James Dean in 1955. Dean was wearing his belt and died instantly, while his unbelted passenger was thrown clear of the car and survived. Yet repeated studies show that seatbelts do save lives – an estimated 1,400 annually in the UK, and as many as 13,000 in the USA. In total, they are reckoned to have saved over a million lives since 1959.

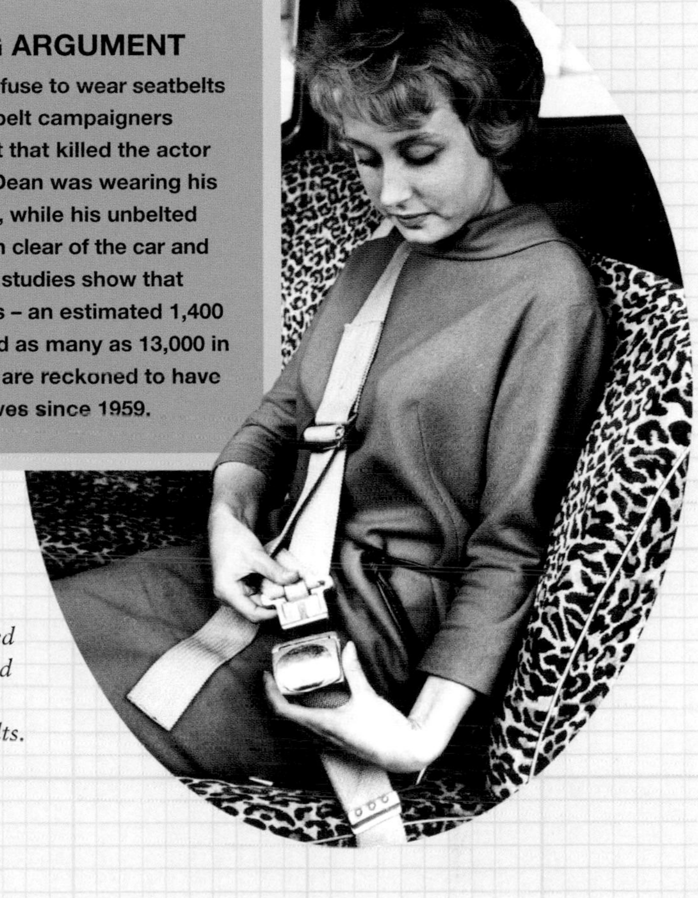

Clunk, click
From 1967, every vehicle in Britain had to have points installed on the door pillars and floor for the eventual fitting of front seatbelts.

Hood hairdryers c1960

In 1960 Calor introduced a hood hairdryer – a domed device that enclosed the top of the head and was ideal for creating the perms then in fashion. The principle was a simple one: electric elements in the hood heated a constant airflow from a fan. The contraptions became a common sight in ladies' hairdressing salons in the 1960s, with rows of women seated under them reading magazines. One of the most iconic hairdos of the early 1960s, the 'beehive', was created by winding a client's wet hair around rollers, drying it under a hooded dryer until set, then combing and reinforcing the style with hairspray.

Starting young
A child emulates her older neighbours in this hairdressing salon in the 1960s.

The rise and rise of the box

Television took off in the West in the decades after the Second World War. Beginning with just a few gargantuan black-and-white sets with tiny screens humming away in the sitting rooms of the affluent, by the start of the 21st century there were 3.5 billion TV viewers worldwide and hundreds of different privately owned networks.

De Gaulle in colour
General de Gaulle, president of France from 1959 to 1969, grasped the importance of putting across a positive public image in the age of television. He made no fewer than 81 broadcasts during his 11-year term.

LIVE FROM THE ABBEY

A defining moment in British television broadcasting history came with the coronation of Queen Elizabeth II on 2 June, 1953. This was the first time that cameras had been allowed inside Westminster Abbey and, with 56 per cent of the population gathered round TV sets to watch, this could be said to be the birth of television as a mass medium in the UK. It took the BBC a year to organise the broadcast, with a team of 120 people and 21 cameras. The broadcast was also beamed live to France, Holland, Belgium and West Germany, so 20 million people across Europe also watched the display of pomp and pageantry.

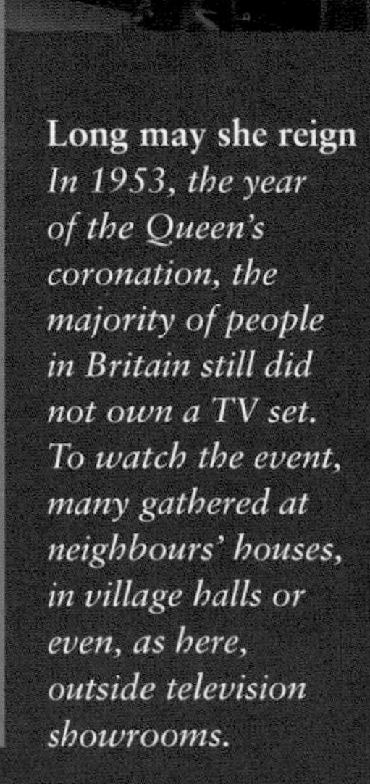

Long may she reign
In 1953, the year of the Queen's coronation, the majority of people in Britain still did not own a TV set. To watch the event, many gathered at neighbours' houses, in village halls or even, as here, outside television showrooms.

After the war, TV stations mushroomed in the United States. From just 17 in 1948, there were 97 in 1950 and 580 a decade later. In line with this, households owning a set increased from 9 per cent in 1950 to 87 per cent in 1960. In Britain, where Scottish inventor John Logie Baird had created the first flickering television images back in the 1920s, broadcasting services resumed in 1946, having been suspended during the war. The BBC had a broadcasting monopoly until the launch of ITV in 1955. The number of TV sets in Britain rose from just 1.5 million in 1952 to close on 7 million in 1957, while that same year, even in post-war West Germany, the figure reached 1.2 million.

Elsewhere in Europe, television was slower to develop. From late 1949, France's only broadcaster started transmitting 16–20 hours of programming a week, exclusively in and around Paris. By 1954, just 1 per cent of French households had a set; it would be 1970 before the television found its way into the majority – 80 per cent – of French homes. The Italian state-run station RAI (Radio Audizioni Italia) began television broadcasting in 1952; Sweden and Spain followed suit in 1956. One of the last developed countries to embrace television was South Africa, where the apartheid regime allowed a strictly controlled nationwide service to begin in 1976.

Information first

The founder of the BBC, Lord John Reith, defined its mission as being to 'inform, educate and entertain'. In European countries in the 1950s the emphasis was strongly on the first of these elements, with television considered to be a public service and charged with promoting

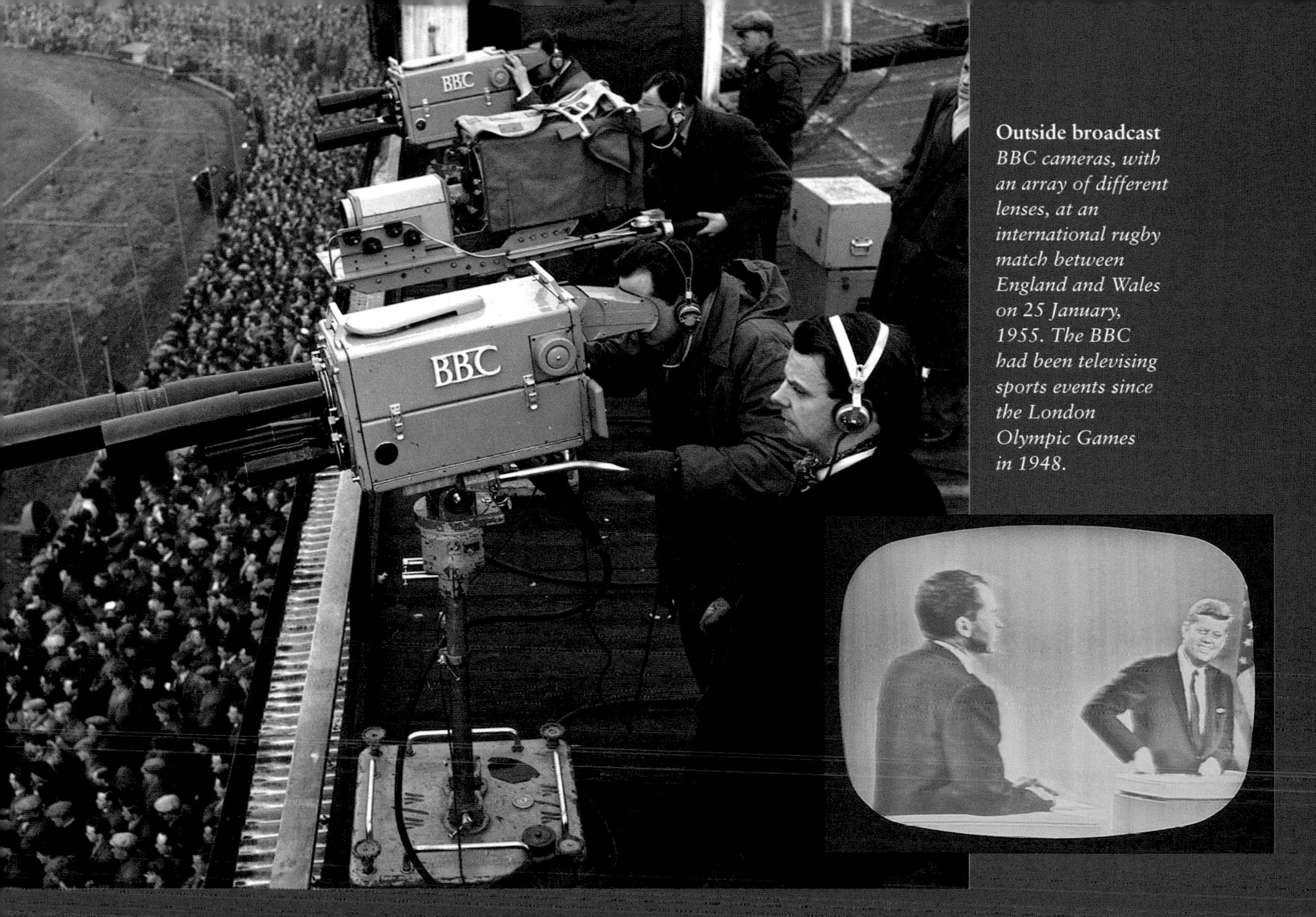

Outside broadcast
BBC cameras, with an array of different lenses, at an international rugby match between England and Wales on 25 January, 1955. The BBC had been televising sports events since the London Olympic Games in 1948.

A television face
On 21 October, 1960, US Senator John F Kennedy and Vice-President Richard Nixon met for the last of four televised debates in the presidential race. A poll afterwards showed that radio listeners thought Nixon had won the debate, but TV viewers were in no doubt that the victor was the telegenic young senator from Boston.

national culture. In 1948 the American network CBS made television history when it introduced the first newscast, a 15-minute daily bulletin entitled *Douglas Edwards with the News*. European broadcasters thought the presence of a newsreader with visible facial movements would distract viewers from the serious subject matter, so at first presenters were hardly seen on screen and an off-camera voice commented on still images. The first newscasters on BBC television, including Kenneth Kendall and Robert Dougall, appeared on 4 September, 1955, in full evening dress.

As television started to become established in Europe, questions arose concerning the relationship between the state and the broadcast media and the independence and impartiality of newsgathering. While democracies were far removed from authoritarian regimes where television became the mouthpiece of a one-party state, the authorities were still keen to keep a close eye on the content of programmes. Politicians either chose to remain aloof – Winston Churchill did not give a single television interview during his entire final term as Prime Minister in the 1950s – or actively intervened to prevent material from being aired that they considered against the national interest. It took some while for even those politicians who were seasoned public speakers to master the new medium. But over time, television became a key element in how the modern politician connected with voters. This point became abundantly clear when, on 26 September, 1960, 71 million Americans tuned in to watch the first televised debate between presidential candidates John F Kennedy and Richard Nixon. A tanned and relaxed Kennedy trounced his ill-at-ease rival in the debate – and went on to do so in the election.

Educate and entertain

From its earliest days, television regarded education as an important part of its remit. In contrast to the primacy that American networks placed on ratings, European broadcasters saw television's public-service ethos fulfilled by educational programmes aimed at both children and adults. The BBC began transmitting the children's puppet show *Muffin the Mule* in 1946, followed in 1952 by the cycle of programmes for pre-schoolers called *Watch With Mother*, including the characters Andy Pandy and Bill and Ben, the Flowerpot Men. In the United States, the

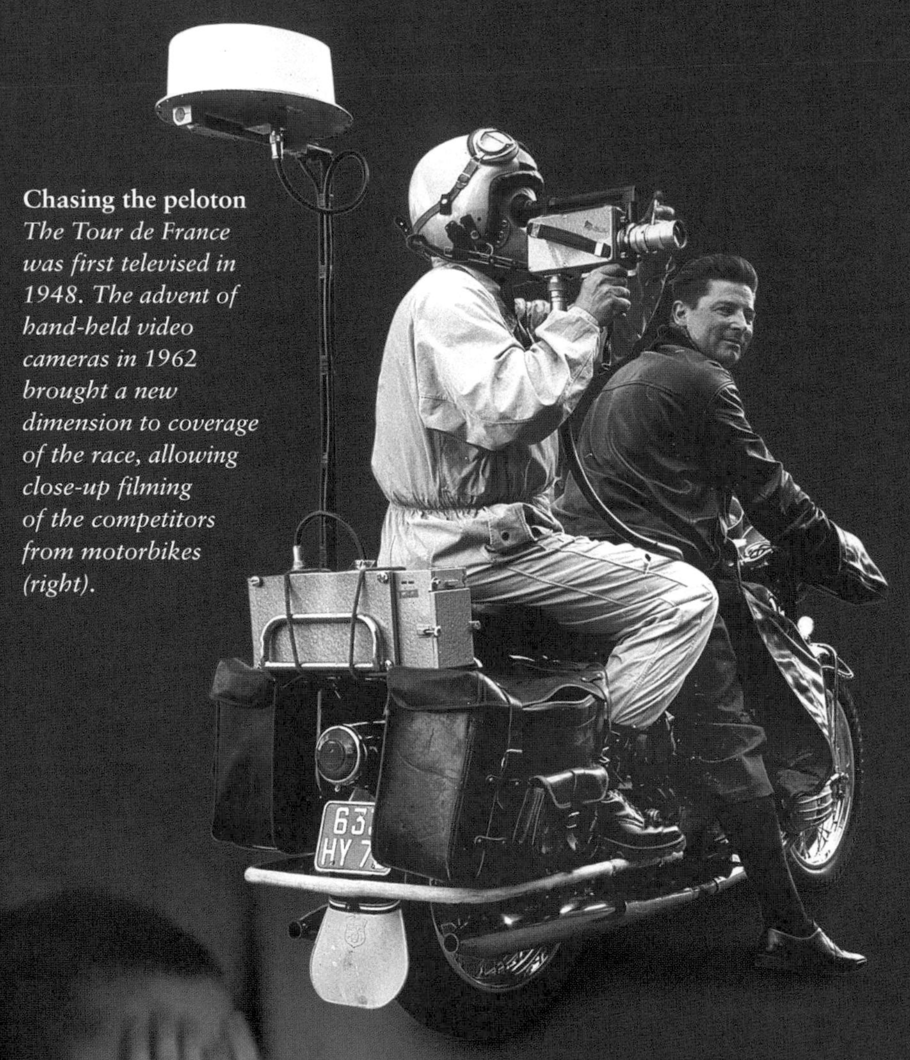

Chasing the peloton
The Tour de France was first televised in 1948. The advent of hand-held video cameras in 1962 brought a new dimension to coverage of the race, allowing close-up filming of the competitors from motorbikes (right).

puppet series *Sesame Street,* much praised for its blend of education and entertainment, first came on air in 1969. For adults in the UK, the newly founded Open University started broadcasting its distance-learning courses on television in 1971.

Television, it was hoped, would also help to foster international understanding. In 1954 the Eurovision Network was founded in Geneva for the purpose of exchanging programmes between Great Britain, France, Holland, Belgium, West Germany, Switzerland and Italy. Its most famous broadcast is the annual *Eurovision Song Contest*.

Sports fans were also well served from the outset. From the comfort of their living rooms, they could watch every twist and turn of a rugby or football match, in far better detail than if they were at the actual ground. Television screening rights now make up a major part of the funding for internationally popular sports such as football and golf.

European TV companies also invested a great deal in developing entertainment programmes. Popular British shows included *Sunday Night at the London Palladium*, *The Billy Cotton Band Show* and *The Black*

Hitting the spot
From 1959 to 1963, American prime-time TV audiences were gripped by The Untouchables, *a series starring Robert Stack as FBI agent Eliot Ness (right). The hugely successful show was syndicated to many other countries.*

GENERATION TELEVISION

In the half-century of its existence as a mass medium, TV has had a profound impact on the way children interact with the world. Television suddenly confronted children with the complexities of adult life, including such controversial areas as violence and sex. Surveys have found that the average American child spends more than 25 hours a week in front of the television. Paediatricians advise that, in order to learn properly how to socialise with others, children should not be exposed to TV before the age of 2.

'*Eagle* has landed'
An estimated half a billion people watched the live broadcast of men landing on the Moon on 21 July, 1969. The first walk on the Moon's surface by Neil Armstrong and Buzz Aldrin is widely considered to be the most memorable event ever shown on television.

ADVERTISING OVERLOAD

Manufacturers and other business interests quickly spotted the selling power of television, and US broadcasts were soon inundated with advertisements. The trend has continued to grow to the present day – over an average 10 hours of TV viewing in America, 3 hours will be filled by commercials. In other countries the frequency of advertising is limited by statute – the UK maximum is 12 minutes per hour, with an overall average of 7 minutes.

and White Minstrel Show, Spain had *Club Miramar*, Italy *Un Due Tre*, while France produced *La Grande Farandole*. To supplement these home-grown programmes, successful series were bought in from America, notably *Bonanza*, *Wanted Dead or Alive* and *The Untouchables*. Game shows, pioneered in the USA by programmes such as *The $64,000 Question* (1955–58), also proved a hit and continue to attract large audiences today.

A changing world

Commercial television developed as a result of independent rather than national broadcasting. The first TV advertisement was aired in the USA on 1 July, 1941, when the watchmaker Bulova paid $9 to have an advertisement screened before a baseball game between the Brooklyn Dodgers and the Philadelphia Pillies. The first advertisement in Britain was for Gibbs SR toothpaste, aired on the first day of independent transmission from the Guildhall in London on 21 September, 1955.

Television gained a further momentum with the introduction of colour, which began in the USA in the mid-1950s, spreading to Europe by the late 1960s. Three different encoding systems were adopted: NTSC (in the USA, Japan, and most Central and South American countries), PAL (in the UK, Germany, Spain, Italy and Scandinavia) and SECAM (France, Russia and Francophone Africa).

Two momentous events of the 1960s – the assassination of President Kennedy in Dallas, Texas, in November 1963 and the first landing of men on the Moon in July 1969 – were broadcast live. They have since become the most memorable TV broadcasts of all time.

The 1980s brought two important developments for the future of television: the arrival of 24-hour news broadcasting and the expansion of commercial television. Both came about as the result of the increased availability of television signals offered by satellite and cable technology. The Cable News Network (CNN) was founded in 1980 in the United

States by entrepreneur Ted Turner. As the first all-news channel it came to particular prominence during the First Gulf War of 1991, with exclusive reports from Kuwait and Iraq as the US-led Coalition expelled Saddam Hussein's forces from Kuwait. The rolling news coverage established a pattern that has been widely imitated. Satellite broadcasting brought an explosion in the number of commercial channels, many of them owned by powerful businessmen with widespread media interests. News International, owned by Australian tycoon Rupert Murdoch, owns television networks in Italy, New Zealand and Croatia, as well as the UK. Murdoch's dominance of the UK media has raised concerns about influence over voter behaviour and political decision-making. When Prime Minister John Major won a surprise fourth election victory for the Conservatives in 1992, for example, one of Murdoch's tabloids declared: 'Its The Sun Wot Won It', and more recently his media backed Prime Minister Tony Blair's commitment of forces to the Iraq War in the face of huge public opposition. An even clearer case of political power amassed through broadcasting is seen in Italy, where entrepreneur Silvio Berlusconi, who broke the state TV monopoly in the 1980s, used his television empire to promote his prime ministerial campaigns.

End of an era
Television news crews were present en masse to witness the German public dismantling the Berlin Wall following the collapse of the communist regime in East Germany on 9 November, 1989.

THE EFFECT ON THE CINEMA

Media analysts have often predicted the demise of cinema in the face of competition from television. The boom in television ownership in the 1960s brought predictions of cinema doom, then again with the introduction of video players in the early 1980s and in the decade following the rise of the internet. Yet after an initial decline in box-office takings, cinemas have found ways of attracting audiences back to the big screen. The ability to save TV shows on digital recorders for later viewing, or to watch on demand on laptops or PCs, has helped cinema to retain a healthy market share; cinemas in Europe, for instance, still attract about a billion film-goers annually. Some TV companies, such as HBO in the United States and the BBC and Channel 4 in the UK, are also actively involved in funding and making new big-screen films.

Rolling news
The American CBS network began its regular evening news broadcasts in 1948, hosted by Douglas Edwards (above). In 1980 (right) it was the turn of the cable broadcaster CNN to break the mould, when it introduced 24-hour newscasting.

Global reach

Television today reaches virtually every corner of the planet. Here, a group of Hmong people – an ethnic group living in the mountainous regions of Vietnam, Thailand and Laos – watch a communal television set in their village (above).

A bewildering array

From the 1990s on, the continuing deregulation of public service broadcasting and explosion of new media have fundamentally changed the ways in which television is produced, distributed and consumed. The choice of channels is now wider than ever before, with free-to-air broadcasts being supplemented by subscription-only viewing available as part of satellite channel packages. Digital signals and the convergence of technology means that people can now watch television on different platforms, such as laptops or 3G mobile phones, or record programmes for later viewing on digital recorders or portable media players.

Recent trends have seen inexpensive reality TV, soaps and talent shows come to dominate the all-important ratings, with a corresponding decline in high-investment documentary film-making and original drama. Television companies now increasingly outsource programme-making to independent production companies, while budding documentary and film-makers are finding new outlets for original material, such as uploading to internet video-sharing sites such as YouTube. In the early 21st century television has become just one of many visual media available, and has changed almost beyond recognition from its inception around 60 years ago.

DIGITAL BRITAIN

The analogue television signal is being phased out in Britain over a period of four years, from 2008 to 2012. In mid-2010, the media watchdog OFCOM reported that 92 per cent of households now owned a digital television and that 71 per cent of secondary sets had been converted with the addition of a digital box, so 80 per cent of all sets in the UK were digital-enabled.

Exposing new talent

In 2009 an unknown filmmaker from Uruguay named Federico Alvarez uploaded onto You Tube a short science-fiction video called Panic Attack. *He had made the film on a budget of just $500, yet it secured him a $30 million deal.*

Bringing coherent light to bear

As early as 1917, Albert Einstein speculated on the possibility of stimulating atoms and molecules so that they would emit light. In 1960 the idea became reality: American physicist Theodore Maiman announced that he had built the world's first laser, a revolutionary device that could supply a beam of light of a single wavelength.

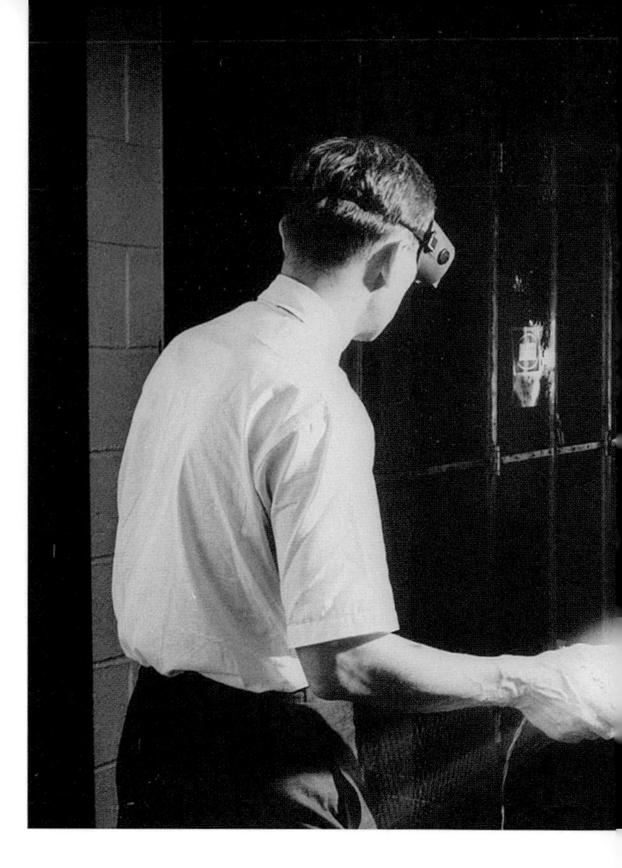

Test firing
Technicians at the RCA Laboratories at Princeton, New Jersey, testing an argon laser in 1967 (right).

The article announcing the invention of the laser contained just three diagrams and was barely 250 words long. Signed by Theodore Harold Maiman, a physicist at the Hughes Research Laboratories in Malibu, California, it was published in the journal *Nature* on 6 August, 1960. Almost 25 years later, reflecting on his groundbreaking work, Maiman recalled that the budget for nine months of research by his team on the laser was a modest $50,000. Today, so ubiquitous have lasers become in many walks of life that standard devices can be manufactured for just a few pence per unit.

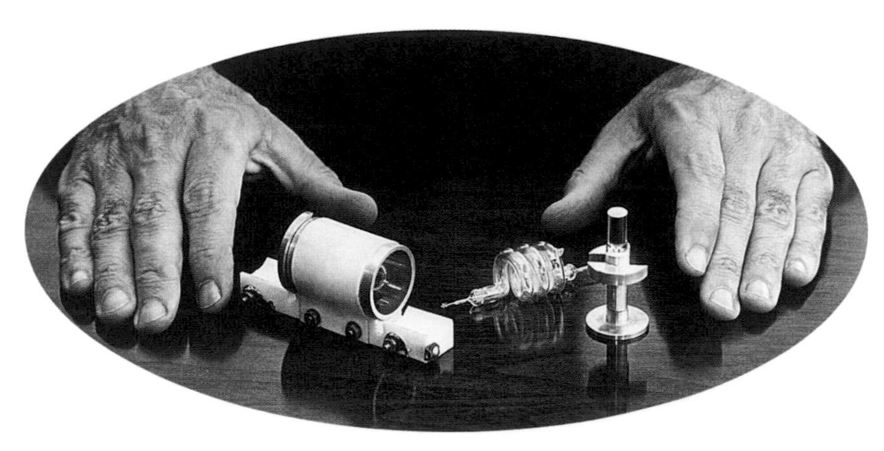

Let there be light
Maiman's laser prototype of 1960 consisted of a cylinder (on the left) and a helicoidal flash-bulb surrounding a crystal of synthetic ruby.

Organising light

The acronym LASER stands for Light Amplification by Stimulated Emission of Radiation, but how does one actually work? Perhaps a helpful analogy is to imagine ordinary light as a random, milling crowd of people, while the light produced by a laser resembles a military parade. The light particles involved are all of a single wavelength and therefore colour, and all the waves are in step. This is what physicists call 'coherent light'. By comparison, the photons in ordinary light radiate in all directions from their source and arrive in totally unsynchronised fashion at their destination – that is, they have neither spatial nor temporal coherence. Because a laser does have this dual coherence, a laser beam can be concentrated on very specific areas or alternatively transmitted over long distances without any loss of power. This means that it can deliver large bursts of energy or quantities of data ultra-fast.

THE LASER EFFECT

A laser gun is rather like a swimming pool with an artificial wave machine, in the form of a large paddle mounted at one end. The paddle, which can move back and forth, creates an initial wave that travels down the pool to the far end, strikes the side and returns. If the paddle strokes are calibrated so that it makes a second wave at precisely the moment the first returning wave hits the paddle, this will generate a superimposition: in other words, the two waves will combine to form a single wave twice the height of the first one. In turn, this wave will travel down the pool, rebound and hit the paddle again, and so on, all the while increasing in amplitude. After n oscillations the wave, now the height of n wavelets, will overflow the pool and flood the surrounding lawn. In terms of light, this corresponds to the emission of a laser beam.

One huge wave

Light exists both as a wave and as particle, and the extraordinary coherence of laser beams is easy to understand if we consider light's wave properties. Looked at in this way, the laser is an enormous light wave formed by superimposing billions upon billions of smaller individual waves. In concrete terms, the laser effect is produced within an atmosphere full of stimulated atoms emitting light waves that built up into a single mega-wave. This mega-wave behaves in exactly the same way as an individual wave, vibrating at a single wavelength and spreading out in a straight line. If it were possible, this would be tantamount to taking every human being on the planet and forming a single person from them, eight billion times the girth of the average individual. The phenomenon of superimposition is commonplace in the realm of wave-particles (even though the laser effect rarely occurs spontaneously). The laws of quantum physics envisage superimposition as a theoretical possibility, but in lasers it has been necessary to induce it artificially.

Powerful apparatus
A close-up of the central component of the world's first TEA (transversely excited atmospheric) pulsed laser, which was designed at the Royal Signals and Radar Establishment in Worcester in 1973. Laser light is generated within the cylindrical chamber, which is filled with carbon dioxide gas as a gain medium.

THE OPTICAL CAVITY

A vital component in generating a laser beam is a pair of mirrors, set parallel to one another at either end of the substance whose atoms are being stimulated (the so-called 'gain medium' or 'lasing medium'). The space between the mirrors is known as the optical cavity. As the gain medium is exposed to an external energy source (a sudden intense flash of light or heat) photons bounce back and forth through it, in the process stimulating the emission of more photons of the same wavelength and phase (direction) as the light shining on it. The mirror at one end of the laser is fully reflective, while that at the other end is half-silvered, so that it reflects some light but lets some light through. When the energy within the optical cavity has gained sufficient power, it breaks through the partially silvered mirror in the form of a concentrated laser beam.

Einstein vindicated

In a 1917 paper entitled 'On the Quantum Theory of Radiation', Einstein foresaw the possibility of stimulating such a superimposition of waves of electromagnetic radiation. Yet more than four decades were to elapse before this became a reality. The gap between theory and practice rested on the matter of how to synchronise the emission of wave-particles of light by billions and billions of atoms. Although various experiment bore out Einstein's prediction at the atomic level – Ladenburg in 1928, Lamb in 1947 – it was only in 1950 that the French physicist Alfred Kastler devised a method for exciting a large

Image from Cassini
Saturn in August 2009, from a distance of 847,000km

A NATURAL LASER

The laser effect can occur spontaneously if a mass of atoms is suddenly subjected to an intensely powerful electromagnetic field. This happens, though rarely, in some envelopes surrounding stars, where the gases are subjected to intense radiation, stimulating a reflection back to the radiation source. Such an occurrence was captured by the *Cassini* space probe on 23 July, 2006, when it photographed a laser-type emission coming from the A and B rings of Saturn. This had been caused by a combination of light rays from the Sun and the effect of electromagnetic fields generated by the planet.

POPULATION INVERSION

In his paper of 1917, Einstein noted that, in order to produce an amplification of light, the optical cavity must always contain more excited-state atoms (those with higher-energy electrons) than those in a non-excited state. Thus, when an excited-state atom emits a photon, this can be reabsorbed by a neighbouring non-excited atom, which in turn becomes excited and emits photons. The phenomenon by which the number of atoms in an excited state comes to exceed those in a ground state (which have some 2 to 3 times less energy) is known as 'population inversion'. This is a vital step in the functioning of a standard laser.

Pinpoint accuracy
Laser positioning systems like this (right) enable technicians to measure distances using a long-range monochromatic gas laser (a mix of helium and neon). These systems emit a beam that diverges by less than one part in a thousand, which is why they are favoured by the military for guiding missiles to their target. Deployment at night is most effective.

number of atoms simultaneously. This process, called 'optical pumping', represented a significant theoretical and technological advance, and won its inventor the 1966 Nobel prize for physics. Kastler's work enabled other scientists to create the first MASERs ('Microwave Amplification by Stimulated Emission of Radiation'), devices that generated electromagnetic waves (radio waves) rather than light. In 1953 the first maser, using energised molecules of ammonia gas, was built by a team led by Charles H Townes at Columbia University, New York. It helped to pave the way for the introduction of the laser.

From infrared to X-rays

In the years that followed two teams of scientists – Charles Townes and Arthur Schawlow at Bell Laboratories, and the Russians Aleksandr Prokhorov and Nikolaï Basov – devoted their energies to developing the first laser. These efforts bore fruit in 1958, when Townes and Schawlow constructed an optical cavity from two highly reflective mirrors placed parallel to one another and used it to demonstrate how the principle of the maser could be applied to producing and amplifying visible light. Townes and the two Soviet researchers were awarded the Nobel prize for their pioneering laser work in 1964, while Schawlow had to wait for his until 1981.

The final breakthrough was achieved by Maiman. His optical cavity comprised a cylindrical rod, 10mm in diameter and 20mm deep, made from a single synthetic crystal of ruby. It was silvered at one end, partially silvered at the other, and lit by monochromatic light generated by strobe lamps. The beam was infrared, but since then lasers have been made

Moon measurement
One of the laser reflectors set up on the Moon by the Apollo 11 *astronauts. These devices enabled scientists to measure the distance from the Earth to the Moon by means of a laser beam shot from Earth through a telescope and trained on the reflector. The results are accurate to within a few centimetres*

OPTICAL PUMPING

The technique of optical pumping, which creates the crucial population inversion within the gain medium, was developed by Alfred Kastler in 1949–50. Kastler succeeded in controlling the behaviour of atoms by manipulating the polarisation of the incident light: it is this factor that determines whether atoms absorb energy or re-emit it. Thus, by placing the atoms within the optical cavity in a magnetic field and exciting them with polarised light, he found that he was able to stimulate the laser effect more efficiently.

for all wavelengths of electromagnetic radiation, from infrared light to X-rays. This diversity has led to the adoption of lasers in a wide variety of different fields.

The most common are low-energy infrared lasers, used in digital devices such as computer mice and laser printers. Tens of millions of these are produced annually. This type of laser is also used in barcode readers, which emit a very narrow beam that tracks back and forth across the surface of the code. According

Sound and light
Lasers have become a familiar part of the show at rock concerts and other stadium events. They are often projected onto a backcloth of dry-ice fog to create spectacular effects. The laser show below is from a concert by the electronic-keyboard virtuoso Jean-Michel Jarre at the National Indoor Arena in Birmingham on 24 May, 2009.

Laser vision
The superhuman powers of Superman, the comic-book hero, included laser vision. This enabled him to perform feats like welding together the damaged hull of a tanker to avert disaster, as portrayed on film by Christopher Reeve (left) in Superman III.

Zoning in
Before radiotherapy treatment begins (below), surgeons deploy lasers to pinpoint the area of the patient's body that will be subjected to treatment.

to the sequence and thickness of the bars, a corresponding part of the beam is reflected back to a sensor that analyses and recognises the code. In a laser computer mouse the infrared beam moves with the user's hand, triggering an optical sensor system; this registers how far the mouse has moved by bouncing hundreds of images every second, constantly updating the position of the mouse and the subsequent position of the cursor on the screen. The same holds true for the lasers that store and read information on CDs.

Multiple roles

Lasers have many applications. Infrared thermal lasers are used in medicine, as they can focus intense heat on a very restricted area in extremely short bursts, typically 10^{-12} seconds. This enables them to be used as cutting instruments in microsurgery of the eye – to mend a detached retina, for example – or for the removal of melanomas from the skin.

Temporal coherence means that lasers can generate ultra-fast beams; thus chemical reactions can be photographed, or even slowed down or accelerated, by lasers operating at a rate of some 10^{14} exposures per second (known as a laser femtosecond). This same principle can be applied to measuring distances. By measuring the time that elapses between the emission of a split-second pulse and the arrival of its reflection, for instance, the distance of the target object from the laser can be calculated. This is called lidar ('light radar'), a technique that has been used – through laser reflectors with a minuscule margin of error – to measure the distance of objects as far away as the Moon.

Meanwhile, the spatial coherence of lasers makes them ideal for use in various military

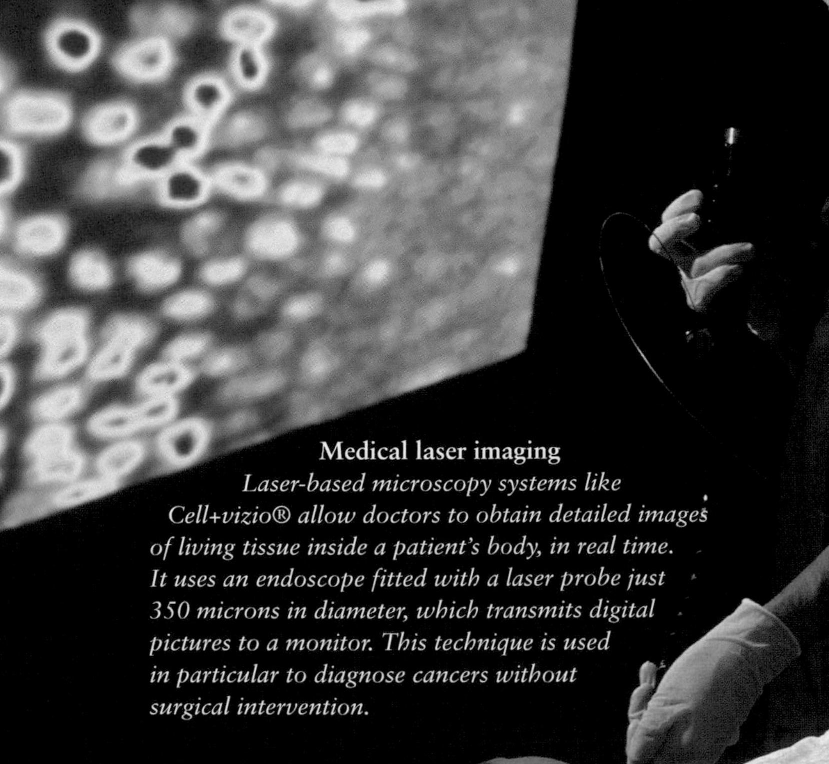

Medical laser imaging
Laser-based microscopy systems like Cell+vizio® allow doctors to obtain detailed images of living tissue inside a patient's body, in real time. It uses an endoscope fitted with a laser probe just 350 microns in diameter, which transmits digital pictures to a monitor. This technique is used in particular to diagnose cancers without surgical intervention.

Like a knife through butter
An industrial laser slices through a block of stainless steel. Aside from its precision accuracy, another advantage to laser cutting is that, since the beam is extremely narrow (typically just 0.5mm for metal), no heat damages the surrounding material. Lasers have many applications in industry, from cutting, drilling and shaping parts to welding them together by melting adjacent edges.

applications, such as in guidance systems for so-called 'smart' bombs. Lasers are also employed to guide machines that lay pipelines and bore tunnels. Another application, which accounts for a third of the civil laser market, is as cutting or welding tools in numerous areas of industrial manufacturing.

Recently, in the USA, experiments have been conducted into harnessing the power of the largest laser ever built, comprising 192 separate beams, to initiate a controlled nuclear fusion reaction. If developed on a commercial scale, this would be an important new source of energy. Meanwhile laser weapons, like the 'phasers' in *Star Trek,* remain firmly in the realms of science fiction.

Last but not least is the role played by lasers in data transfer. Billions of bits of information per second – be it computer files, voice signals in telephone traffic, or digital television channels – can be transmitted by lasers down fibre-optic cables. It is no exaggeration to say that lasers revolutionised telecommunications worldwide.

A LASER FOR EVERY JOB

The huge range of lasers that now exists, varying both in the wavelength (colour) and the power of the light they generate, derives from the fact that the gain medium within the optical cavity may comprise any number of different types of atoms or molecules in every conceivable state – solids, liquids or gases. Most lasers fall into the following three broad categories:

- Gas lasers – including helium-neon, argon, krypton, CO_2, and carbon monoxide types.
- Chemical lasers – hydrogen fluoride, deuterium fluoride and oxygen iodide.
- Solid-state lasers – these use a medium made from a crystalline substance such as ruby or a doped-crystal material such as titanium-doped sapphire. There are also lasers in which the active medium is a semiconductor.

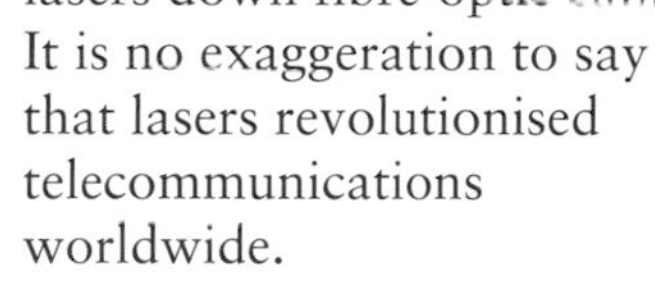

Virtual typing
The projection keyboard is a new device that can be used with bluetooth-enabled cellphones or computers to project the hologram of a full-size keyboard onto any flat, opaque and nonreflective surface.

Home of the EEC and EXPO 58

On Thursday 17 April, 1958, the Brussels World's Fair opened its doors to the public. The significance of the exhibition went well beyond the obvious pride it instilled in the small country of Belgium and its capital. As the first such fair since 1939, it signalled Europe's recovery and emergence from austerity after the ravages of the Second World War.

Bureaucrat city
Designed by the Belgian architect Lucien de Vestel, the Berlaymont building in Brussels was built in 1963–7 to house the 3,000 civil servants of the European Commission. The huge size of the building has earned it the nickname 'Berlaymonster'.

By the late 1950s, Belgium was looking towards the future with confidence. Aided by the US Marshall Plan, its industry had recovered from the damage caused by the war. Its export figures were respectable and unemployment levels negligible – just 110,000 Belgians were out of work, from a total population of more than 8.5 million. Yet there was also another major reason for the country to feel optimistic. West Germany, France, Italy, the Netherlands, Luxembourg and, of course, Belgium itself – the six founder members of the European Economic Community (EEC), formed by the signing of the Treaty of Rome on 25 March, 1957, and inaugurated on 1 January, 1958 – had just chosen Brussels as the seat of the European Commission. Under the guidance of the French prime minister Guy Mollet and his German counterpart, Chancellor Konrad Adenauer, the process of European cooperation and integration, though initially restricted to a customs union, had as its ultimate goal the formation of a strong trading and diplomatic bloc to counterbalance on the one hand the growing power of the United States and on the other the Soviet Union.

Brussels was also the headquarters of Euratom, the organisation responsible for promoting the growth of a European nuclear power generation industry to serve the needs of the six member states. There was universal assent that nuclear energy would be one of the key factors in the new industrial revolution that was just beginning. In 1973 three new members were admitted to the European Community: Great Britain, Ireland and Denmark. After several other rounds of accession in the intervening years, current membership of the community (since renamed the European Union, or EU) stands at 27 countries, with a total population of almost 500 million. The collapse of Russian communism and the accession of many former Soviet republics and Warsaw Pact countries have seen the EU reunite the hopelessly divided Europe that resulted from the Yalta Conference at the end of the war.

Flocking to the fair
Special buses, trains helicopters, even motorcycle rickshaws (right) were laid on to bring visitors to the Brussels Expo.

One huge building site

Although India, Australia and China stayed away from Expo 58, more than 45 nations and 7 international bodies accepted the Belgian government's invitation to attend. In addition, there were many multinational industrial concerns, such as the electrical firm of Philips which commissioned the leading Swiss architect Le Corbusier to design its pavilion – a bold, soaring building in pre-stressed concrete. The building's radical design included a sound installation involving 425 loudspeakers (Edgar Varèse's 'Poème électronique') and films projected onto the internal walls.

The pavilions were erected on a 200-hectare site on the Heysel plateau, 7km northwest of the old city centre, that was transformed into a huge open-air building site. A million cubic metres of earth were shifted to prepare for the exhibition. Reckoning on an average of 150,000 visitors a day, the organisers set in motion major civil engineering projects, particularly to improve the transport infrastructure. A network of access roads was built, linking expressways to the city centre, while the ring roads were upgraded to fast, multilane highways. Well before the Expo, Brussels already had an extensive and efficient public transport system, but this was the age when the car became king. In 1958, Brussels and its suburbs contained 110,00 automobiles, or one for every nine citizens.

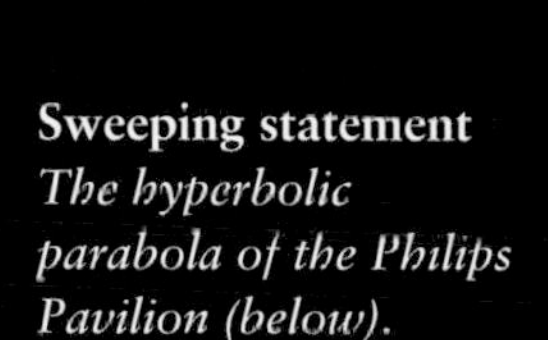

Sweeping statement
The hyperbolic parabola of the Philips Pavilion (below).

MEET AND GREET

To provide information to visitors, take care of any lost children and show VIPs round the site, the Expo 58 organisers recruited 280 young women to act as hostesses. Aged between 18 and 35, they were required to be of pleasant appearance and were kitted out in a uniform comprising a long navy-blue skirt, maroon jacket, white blouse, tie and three-cornered hat. They were forbidden to smoke in public, accept gifts of flowers or chocolate, or to socialise in the evening with any men they met at the Fair.

STAR ATTRACTION

The Atomium (below) – iconic centrepiece of the Brussels World Fair – was designed by André Waterkeyn, director of the Belgian metallurgical federation Fabrimetal. Its nine steel spheres, each 18m in diameter and clad in highly polished aluminium, were arranged in the shape of a cell of iron crystal. Six of the spheres could be accessed by the public, which were connected by 29m-long tubes. The central tube housed a lift that, at 5m/sec, was the fastest in the world at the time, taking 22 people to the top of the building in just 23 seconds. At 35m, the escalators in the connecting tubes were among the longest in Europe. The Atomium was due to be demolished after the Expo, but was retained and has since become the city's top tourist attraction.

Symbol of progress

From the opening of Expo 58 on 17 April to the time it closed on 19 October, Brussels became the centre of the world. The focal point of the Expo was the Atomium, an extraordinary structure standing 102m tall, re-creating an iron-crystal molecule magnified 165 billion times. It was intended as a forward-looking symbol in the Atomic Age.

The Belgian national television service beamed live pictures of the opening ceremony over the Eurovision network. At 11.00am sharp, King Baudouin I, who had come to the Belgian throne in 1951, inaugurated the Fair with a speech in which he urged the world's nations to embrace peace and scientific progress. The theme of the fair was 'A World View – A New Humanism'. Held at the height of the Cold War, Expo 58's aim was to promote international rapprochement through a shared vision of a brighter future, in which people's lives would be made easier by technology.

At the time Britain was taking a lead in radio astronomy and jet-powered flight. The faith in future technology had also been evident at Britain's own giant exhibition, the Festival of Britain, held in 1951. Though still playing a significant world role, Britain was not yet part of the European Union and her ambivalence was echoed in the official British site in Brussels. The displays were divided between the worlds of heritage, as represented in the Hall of Tradition; of scientific innovation, as displayed in the Hall of Technology; and of economics and industrial competitiveness, represented architecturally by the contemporary design of the glass-fronted British Industries Pavilion.

Cold War cultural rivalries

Every participating country tried its utmost, naturally, to present itself in a favourable light. Nowhere was this more evident than in the Soviet Union pavilion, which proudly displayed oversized models of its first two satellites, Sputnik I and II. Loudspeakers played the famous bleeping signal that was emitted by the first Sputnik from 4 October, 1957. These exhibits proved a great draw. The Russians consciously used their high-tech pavilion as an opportunity to persuade fair-goers that a technologically and scientifically advanced USSR would soon outstrip the United States in the production of material goods.

Just a few metres away, the huge circular United States pavilion promoted American values a little less heavy-handedly, showing the

Soviet satellite
A replica of the 'Sputnik' satellite (right) on display in the pavilion of the Soviet Union at Expo 58. Just two years after brutally crushing the Hungarian uprising, Russia was trying to present a positive, progressive image.

wonders of colour television (a recent invention). On the IBM stand, the Ramac 305 caused a sensation: this machine, as large as a wardrobe and weighing a tonne, was the first commercial computer to incorporate a hard-disk drive and could process data at unprecedented speeds.

The Czech Pavilion, a masterpiece of clean, modern design, was visited by more than 6 million people, and was officially named best pavilion of the fair. The pavilions of West Germany, Finland, Austria, Norway, Portugal, Brazil and Mexico were also notable for their bold modern architecture.

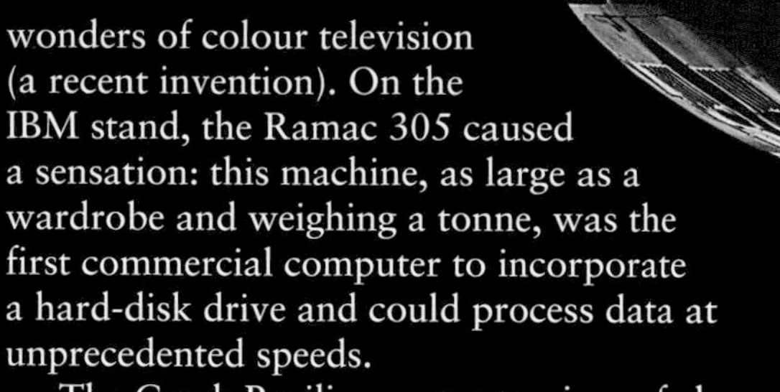

Sheer panache
Above right: the French pavilion, designed by Guillaume Gillet and René Sarger, was an innovative piece of engineering. One exhibit within it was a scale model of the new Citroën DS19 (above), an innovative design that was the work of Flaminio Bertoni.

FACTS AND FIGURES

The construction of Expo 58 involved almost 15,000 labourers, who put in a total of 60 million man-hours. Over the six months it was open, it attracted just short of 41.5 million visitors, including 4 out of every 5 Belgian citizens. Two thousand journalists covered the event. Special transport to ferry people between the 11 separate areas of the Fair included 25 'Expo-trains' (a rubber-tyred carriage pulled by Mercedes-Benz minibuses), a cable-car with 165 cabins and a fleet of 300 motorbike rickshaws.

Age of the peaceful atom

A funfair with space-age-themed rides provided recreation in Heysel Park. For the technically minded, the International Hall of Science had examples of current research into the atom, cells and space exploration. A star exhibit was one of the first super-accurate atomic clocks.

The official poster for the Hall of Science was a striking abstract of different coloured electrons orbiting a nucleus. Just 13 years after the destructive force of splitting the atom had been released on Hiroshima and Nagasaki, Expo 58 was concerned to stress the peaceful potential of nuclear power. It was an optimistic statement of confidence in humanity's ability to cooperate under a banner of progress.

Old-world charm
The Grand Place in Brussels (below), with buildings such as the 16th-century Maison du Roi, was named a UNESCO World Heritage Site in 1998. The ultramodern 1960s quarter housing the European institutions is nearby.

CIRCUIT
CARRY-OVER CIRCUIT
PULSE STANDARDIZER

CHRONOLOGY

The timeline on the following pages outlines key discoveries and inventions from the Second World War to the mid 1960s. Selected historical landmarks are given at the start of each period to provide chronological context for the scientific, technological and other innovations listed below them.

1938

EVENTS

- Nazi Germany annexes Austria in the Anschluss (1938)
- The Munich peace accords signed by Germany, France, Great Britain and Italy enable Germany to annexe the Sudetenland from Czechoslovakia (1938)
- King Carol II of Romania annuls parliamentary democracy and establishes a dictatorship (1938)
- Anti-Jewish violence is unleashed in 'Kristallnacht' (1938)

INVENTIONS

- The KdF-Wagen, prototype of the Volkswagen Beetle, is unveiled in Germany as an affordable family car
- Action Comics, an American monthly cartoon magazine, publishes the first Superman strips by Jerry Siegel and Joe Schuster
- A living coelacanth, a species of fish believed to have been extinct for 70 million years, is caught by a fisherman off the coast of South Africa
- The US textile firm DuPont lodges a patent for nylon, a synthetic fibre developed by the chemist Wallace Carothers
- The world's first aircraft with a pressurised cockpit is tested successfully in the USA
- Indiana University professor Rolla N Harger invents the 'Drunkometer', a breathalyser for testing motorists
- Swiss food company Nestlé introduces Nescafé, instant coffee

1939

EVENTS

- Nazi Germany and Soviet Russia sign a Non-Aggression Pact (1939)
- The German invasion of Poland on 1 September forces Britain and France to declare war on Germany (1939)
- Allies retreat from Dunkirk; Battle of Britain (1940)
- The USA enters the Second World War after Japanese aircraft bomb the American naval base at Pearl Harbor, Hawaii (1941)

INVENTIONS

- Paul Hermann Müller, an employee of the Swiss chemical company Geigy, discovers the insecticide properties of DDT
- The first successful flight by a jet is made in Germany by a Heinkel He178 experimental aircraft, powered by an He-S3 turbojet engine
- Frédéric Joliot-Curie, Hans Halban and Lev Kowarski file a patent describing the theoretical working principles of the atomic bomb
- The Lascaux Caves are discovered in the Dordogne region central France, containing prehistoric painting of animals and hunters dating back to between 15,000 and 18,600 BC
- British mathematician Alan Turing describes the functions of the computer in his abstract theoretical construct of the 'Turing machine'; his work on decryption during the war enables the Allies to break German military codes

► A circuit board from the British ATLAS computer of 1962

▲ A 1958 advert for the French kitchen roll Sopalin

◄ IBM's 'Stretch' computer in 1959

1943

- The Soviet Red Army defeats the German Eighth Army in the Battle of Stalingrad (1943)
- Destruction of the Warsaw Ghetto by Nazi troops (1943)
- Allied forces establish a bridgehead in Normandy in the D-Day landings (1944)
- Liberation of Paris (1944)

- American scientists identify the existence of the jetstream, high-altitude air currents that flow above the Earth
- Jacques-Yves Cousteau and Émile Gagneau invent the Aqua Lung, which allows divers to operate underwater without an air line from the surface
- Research on animal body temperatures, conducted by the American herpetologists Raymond Bridgman Cowles and Charles Mitchill Bogert, calls into question the traditional binary classification of the animal kingdom into warm-blooded and cold-blooded creatures
- German forces deploy their Vergeltungswaffen ('revenge weapons') against Allied targets – the V-1 is the world's first cruise missile, the V-2 the first guided ballistic missile
- American Percy Spencer invents the microwave oven

1945

- Fall of Berlin (1945)
- Atomic bombs dropped on Hiroshima and Nagasaki (1945)
- End of the Second World War (1945)
- Founding of the United Nations (1945)
- Nazi war leaders are tried and convicted of crimes against humanity in the Nuremberg Trials convened by the victorious Allies (1946)

- University of Pennsylvania publishes a report on the construction of a variable automatic calculating machine, a device that soon becomes commonly known as the computer
- In Italy, the former aircraft manufacturer Piaggio lodges a patent for the Vespa, the first modern motor-scooter
- American Earl Tupper invents Tupperware, a range of light and durable kitchen containers made from plastic
- American Vincent Schaefer is the first person to create artificial snow
- Kitchen roll under the brand-name Sopalin, is introduced in France

▼ Actress Audrey Hepburn on a Vespa scooter in the film *Roman Holiday*

▲ The 'Atomic', an Italian espresso coffee machine from the 1950s

◄ Tupperware containers

1947

EVENTS

- Inception of the Marshall Plan, the US economic recovery programme for the reconstruction of Western Europe (1947)
- The India Independence Act gives rise to two new sovereign states, India and Pakistan (1947)

INVENTIONS

- Italian bar owner Achille Gaggia invents the espresso machine.
- American chemist Willard Frank Libby introduces radiocarbon dating
- The first synchrotron, an advanced form of particle accelerator, is built in the United States
- Two employees of Bell Labs create the world's first transistor, an essential component in all modern electronics
- Hungarian-British electrical engineer Dennis Gabor pioneers holography, the reconstruction of 3-D images
- First flight using solely an autopilot takes place between Newfoundland and Britain, a distance of 2,500 miles
- American inventor Walter Morrison makes and markets a throwing disc made from Bakelite, the prototype of the Frisbee

1948

EVENTS

- The Soviet-backed Communist Party seizes power in Czechoslovakia (1948)
- Soviet blockade of rail and road links to West Berlin prompts Allies to begin the Berlin Airlift to keep the city supplied (1948–9)
- State of Israel founded (1948)
- Victory of Nationalists in South Africa and start of apartheid system (1948)

INVENTIONS

- American inventor Edwin H Land launches a revolutionary new camera, the Polaroid, which can develop and print photos within the camera body
- The theory that the Universe expanded from a single point in space and time, better known as the Big Bang, becomes the standard model for cosmologists
- Swiss inventor and explorer Auguste Piccard designs and builds the bathyscaphe, a radical new type of manned submersible
- The first McDonald's® fast-food restaurant opens in San Bernardino, California
- The board game Scrabble, devised by the American architect Alfred Mosher Butts, is given its now-familiar name and successfully marketed by the entrepreneur James Bruno
- Husband-and-wife team Valerie and Pat Hunter patent the first disposable nappy in the UK

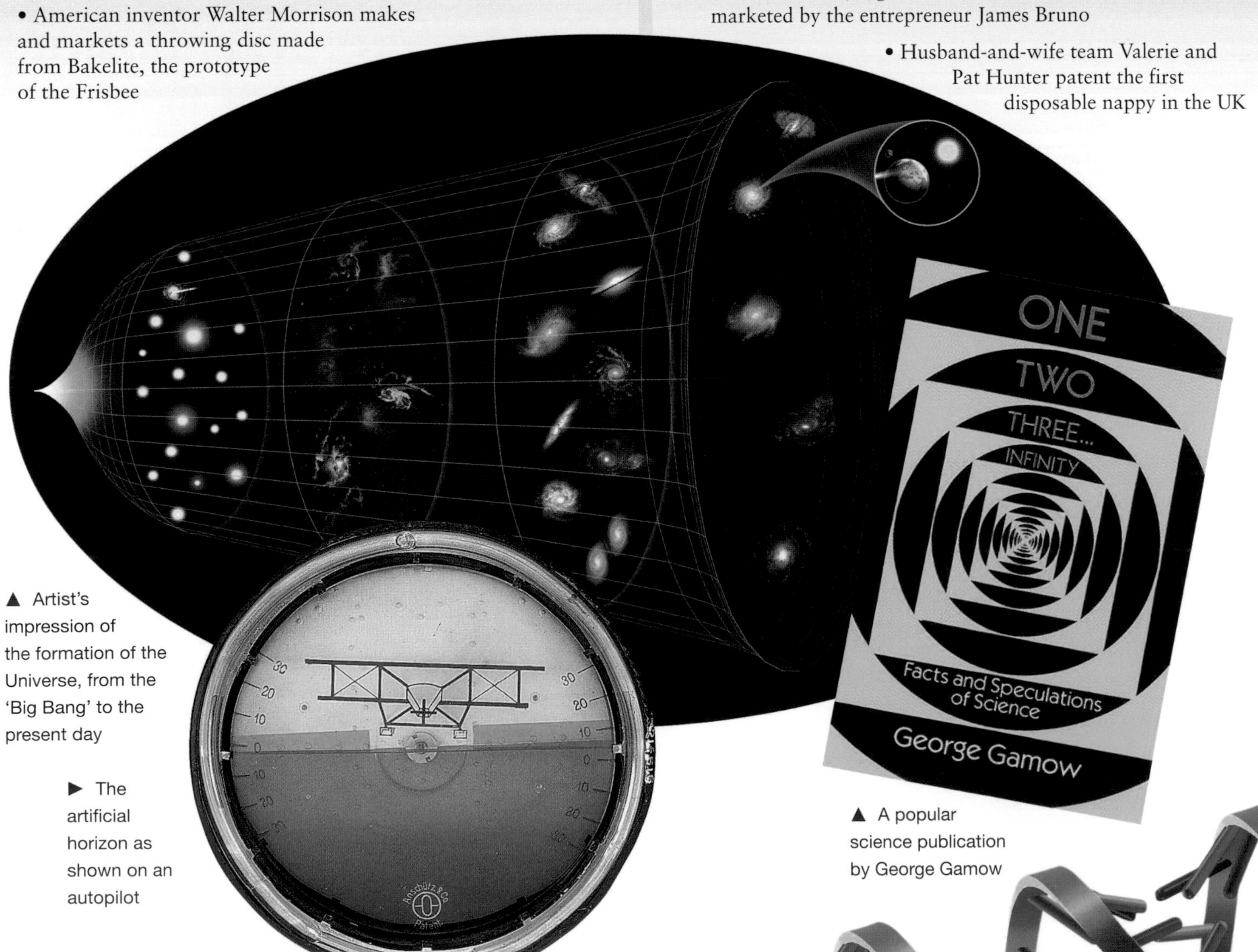

▲ Artist's impression of the formation of the Universe, from the 'Big Bang' to the present day

► The artificial horizon as shown on an autopilot

▲ A popular science publication by George Gamow

1949

- The People's Republic of China is proclaimed in Beijing (1949)
- Formation of the North Atlantic Treaty Organisation (NATO) (1949)
- Invasion of South Korea by the North and outbreak of the Korean War (1950)

- Americans Bernard Silver and Norman Woodland submit a patent for barcodes
- The Comet, the world's first jet airliner, designed by British aircraft manufacturer DeHavilland, makes its maiden flight
- Frank McNamara, head of the Hamilton Credit Corporation in New York, comes up with the idea for the Diners' Club, the world's first credit card
- On 21 November 1951, the research ship *Calypso*, under the command of Jacques Cousteau, sets sail for the first time from Toulon; the vessel becomes Cousteau's base for numerous expeditions studying the marine environment
- First videotape recorder (VTR) using magnetic tape is produced by Armour Research in the USA and further developed by the Ampex corporation
- US physicist William Shockley and others develop the p–n junction transistor, a major step forward in electronics

1952

- Mau-Mau violence breaks out in Kenya (1952)
- Death of Joseph Stalin; Nikita Khrushchev becomes First Secretary of the Communist Party of the Soviet Union (1953)
- The end of the Korean War leaves the country divided along the 38th Parallel (1953)
- Dwight Eisenhower becomes US president; American communists Julius and Ethel Rosenberg are convicted of spying for the Soviet Union and executed (1953)

- Glass ceramics, resistant to high temperatures and sudden shocks, are discovered by the American chemist Donald S Stookey
- Great Britain explodes its first nuclear fission (atomic) bomb
- Canadian physician William G Bigelow introduces the heart pacemaker
- Ultrasound scans, a spin-off from sonar, enable doctors to examine internal organs without invasive surgery
- The first H-bomb is detonated by the Americans on Eniwetok Atoll in the Marshall Islands (South Pacific)
- Television becomes a mass medium in Britain as a result of the BBC's coverage of the coronation of Queen Elizabeth II, watched by a huge domestic audience and also transmitted to parts of Europe
- Francis Crick and James Watson – with invaluable but largely unacknowledged help from Rosalind Franklin – discover the double-helical structure of DNA

▼ Jacques Cousteau's research vessel

▲ A giant barcode on the internet site of the American artist Scott Blake (top)

▲ Model of DNA

1954

EVENTS

• French forces are defeated by the Vietminh at the battle of Dien Bien Phu and lose control of Indochina (Vietnam); Laos and Cambodia also gain independence (1954)
• Formation of the South-east Asian Treaty Organisation (SEATO) (1954)
• Start of the uprising in Algeria against French colonial rule (1954)

INVENTIONS

• Videorecorders are used on a large scale for the first time, by American television networks

• The Soviet AM–1 becomes the first nuclear reactor to generate electricity for civil use

• Frenchman Marc Grégoire invents the non-stick pan (Tefal)

• Researchers Jonas Salk, Pierre Lépine and Albert B Sabin introduce their separate anti-polio vaccines almost simultaneously

• American surgeon Joseph Murray performs the first successful kidney transplant

• The USS *Nautilus*, the world's first operational nuclear-powered submarine, is launched at Groton, Connecticut

• The first vertical take-off and landing (VTOL) aircraft, an ungainly research vehicle called the 'Flying Bedstead,' is developed by Rolls-Royce in Great Britain

1955

EVENTS

• West Germany joins NATO (1955)
• The Soviet Union forms the Warsaw Pact from its Eastern European satellite countries: Albania, Bulgaria, Hungary, Poland, the GDR, Romania and Czechoslovakia (1955)
• The Bandung Conference of non-aligned nations condemns colonialism (1955)

INVENTIONS

• Fibre optics, which can conduct light over long distances, are developed by the Indian-American physicist Narinder Kapany

• British engineer Christopher Cockerell lodges a patent for an air-cushion vehicle – the hovercraft

• Synthetic diamonds are manufactured on an industrial scale

• British scientists Essen and Parry devise the atomic clock, which uses a caesium-beam resonator

• Lego toy building bricks are designed and manufactured by Ole Kirk Christiansen in Denmark

• Swiss engineer George de Mestral invents Velcro, a new form of clothes fastener

• Leading French anthropologist Claude Lévi-Strauss publishes his seminal work *A World on the Wane (Tristes Tropiques)*.

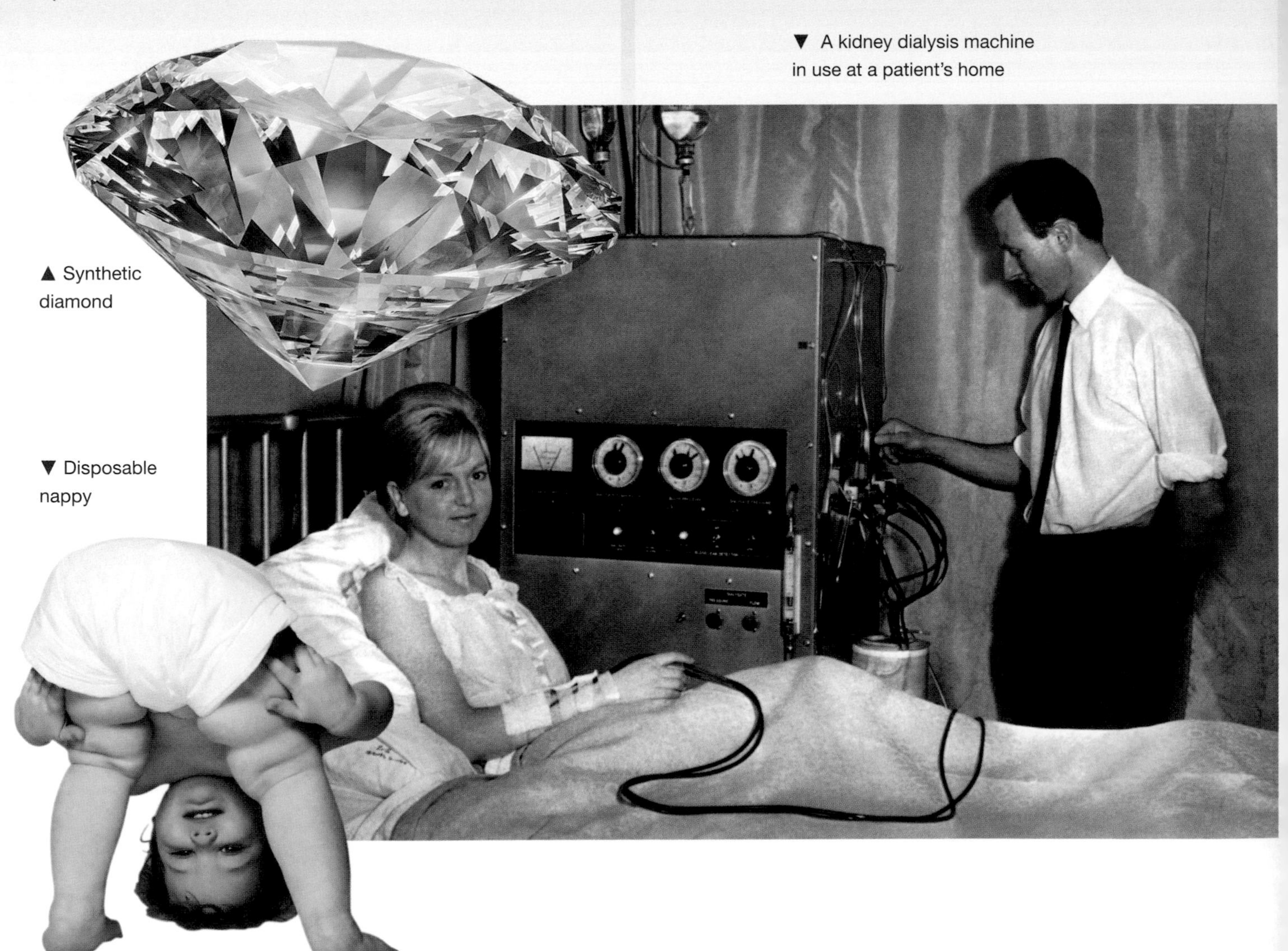

▼ A kidney dialysis machine in use at a patient's home

▲ Synthetic diamond

▼ Disposable nappy

1956

- Soviet Russia embarks on policy of de-Stalinisation and peaceful coexistence with the West (1956)
- Hungarian Uprising crushed by Soviet and Warsaw Pact forces (1956)
- Tunisia and Morocco gain independence (1956)
- President Gamal Abdel Nasser of Egypt nationalises the Suez Canal (1956)

- American Victor Mills introduces Pampers disposable nappies
- Go-karting soars in popularity after American Art Ingels uses a lawnmower engine and scrap parts to build the first powered kart
- The contraceptive pill, developed by Gregory G Pincus using the anti-ovulatory properties of hormones, undergoes successful clinical trials
- Computer-assisted manufacture promotes the automation of factories, revolutionising the manufacturing workplace
- In an attempt to perfect renal dialysis, Dutch physician Willem Kolff invents the artificial kidney.
- The American Mattel toy company launches the Barbie® doll

1957

- First artificial satellite, Sputnik I, is launched by the Soviet Union (1957)
- Chinese Communist leader Mao Zedong inaugurates the 'Great Leap Forward' policy of industrial growth (1958)
- Formation of the European Economic Community (1958)

- International Geophysical Year marks a turning point in polar exploration: over 18 months, beginning in 1958, 67 countries agree to cooperate in studying the climate, geology and fauna of the polar regions, establishing several research stations
- Super Glue, the first cyanoacrylate adhesive, comes on the market
- American inventor Jack St Clair Kilby creates a revolutionary new type of electronic circuit; the integrated circuit
- Expo 58, the first international exhibition since the war, is held in the Belgian capital Brussels
- Canadian Joseph-Armand Bombardier introduces the one-man snowmobile (Ski-Doo)
- The first three-point seatbelt for cars is designed by the Swede Nils Bohlin

◄ Ponytail Barbie® of 1959

► Super Glue

◄ Three-point seatbelt

▼ Pedal-powered karts, the ancestors of go-karts

1960

EVENTS

- Several former British and French colonies in Africa gain independence (1960)
- Building of the Berlin Wall (1961)
- Soviet cosmonaut Yuri Gagarin becomes the first human in space (1961)
- A CIA-sponsored invasion of Cuba is thwarted at the Bay of Pigs fiasco (1961)

INVENTIONS

- The Calor company introduces the hood hairdryer
- American physicist Theodore H Maiman builds the first laser
- Introduction of the halogen lamp
- The General Conference of Weights and Measures adopts the SI system of standard units of measurement, which redefines the metre, kilogram and second
- The audiocassette, pioneered by Philips, begins to challenge the supremacy of the vinyl record
- Kenyan anthropologist Jonathan Leakey discovers remains of the oldest-known human (*Homo habilis*) in the Olduvai Gorge, Tanzania (Rift Valley)
- Laser eye surgery becomes a reality

1962

EVENTS

- War in Algeria comes to an end, heralding independence (1962)
- Stand-off between the USA and USSR in the Cuban Missile Crisis (1962)
- First Civil Rights March in Washington DC (1962)
- Assassination of President John F Kennedy in Dallas, Texas (1963)
- Soviety Union and USA sign Nuclear Test Ban Treaty (1962)

INVENTIONS

- Data transfer between computers down a telephone line becomes possible thanks to the introduction of the RTC modem, precursor of the digital ADSL modem
- Nitinol, an alloy of nickel and titanium, is the first material to have the property of 'shape memory' (superelasticity)
- Death of the pioneering Swiss inventor and explorer Auguste Piccard
- American naturalist Rachel Carson publishes *Silent Spring*, the first book to draw attention to the environmental damage caused by pesticides
- The USA, Great Britain and the Soviet Union agree to stop atmospheric tests of nuclear weapons

► The Philips Pavilion, designed by the architect Le Corbusier, for the Brussels World Fair

▼ A salon hairdryer

▲ Auguste Piccard

1964

• US Congress adopts the Civil Rights Act, paving the way for the end of racial discrimination (1964)
• Dr Martin Luther King is awarded the Nobel peace prize (1964)
• The USA sends troops to Vietnam (1964)

• IBM develops the first computer-aided design (CAD) system and first word-processing program

• Home kidney dialysis is introduced in Britain and the USA

• Development of carbon fibre, a new highly durable material

• Robert Moog creates a music synthesiser that is theoretically capable of producing an infinite variety of sounds

• Container ships are introduced, changing the face of international sea transport and trade

• China tests its first nuclear fission bomb

1965

• Ian Smith declares UDI in Southern Rhodesia (now Zimbabwe) (1965)
• Race riots and the growth of the Black Panther movement in the USA (1966)
• Six-Day War between Israel and neighbouring Arab countries (1967)
• Warsaw Pact forces crush the Prague Spring reform movement in Czechoslovakia (1968)
• Assassination of Dr Martin Luther King (1968)

• The hydroelectric-generating tidal barrier across the estuary of the River Rance in Brittany, France, begins operation; it is the world's largest tidal power electricity plant

• The first portable video recorder and camera are introduced by Sony

• Astronomers Arno Penzias and Robert Wilson detect cosmic microwave background radiation (CMBR), convincing most scientists that the 'Big-bang' theory is correct

• The Mont Blanc Tunnel linking France and Italy is opened to traffic

• The first small, affordable domestic microwave oven is introduced by Raytheon in the USA

• US engineer Ray Milton Dolby develops a method for eliminating background sound on recordings

• South African surgeon Christiaan Barnard performs the first partially successful human heart transplant; the recipient survives for 18 days

• Regular hovercraft services are inaugurated across the English Channel

► Testing an argon laser

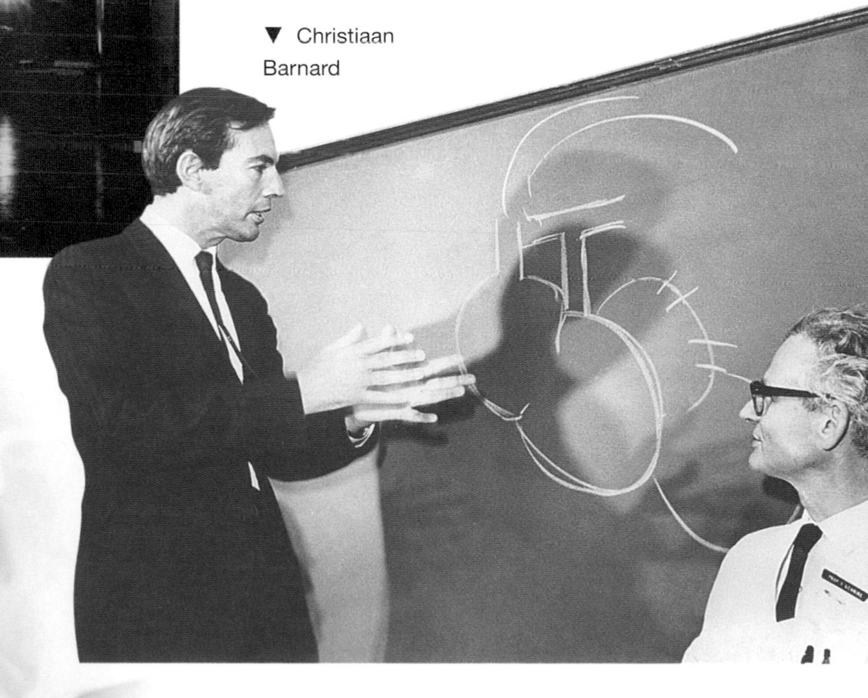

▼ Christiaan Barnard

◄ Use of a laser in radiotherapy

Index

Page numbers in *italics* refer to captions.

Picture credits

ABBREVIATIONS: t = top, c = centre, b = bottom, l = left, r = right

Front cover: main image: aerial view of the Festival of Britain exhibition, held in London in 1951, courtesy of The National Archive UK; inset: a pressure cooker © Groupe Seb.
Spine: Stacking bowls © Tupperware.
Back cover: Printed circuit board, Corbis/Lester Lefkowitz.
Page 2, left to right, top row. © Groupe Seb; Leemage/AGF; The Bridgeman Art Library/DaTo Images, Philips publicity poster, 1952, Pierre Masseau dit Fix-Masseau (1905-1994), private collection © Adagp, Paris 2010; 2nd row: © Tupperware; Cosmos/SSPL/Science Museum; Getty Images/Time & Life Pictures/Co Rentmeester; 3rd row: Corbis/Kulka; Corbis/Ecoscene/Alan Towse; Cosmos/SPL/Science Museum; bottom row: Cosmos/SSPL/NMeM; Cosmos/SSPL/Science Museum; Getty Images/Hulton Archive/Constance Bannister Corp.
Pages 4-5: Leonardo Diaz Romero/Photolibrary.com; 6tr: AKG-Images/SPL; 6cl: Leemage/AGF; 6br: Cosmos/SPL/James King-Holmes; 7tl: Cosmos/SPL/Lawrence Berkeley Laboratory; 7tr: The Bridgeman Art Library/DaTo Images, Philips publicity poster, 1952, Pierre Masseau dit Fix-Masseau (1905-1994) © Adagp, Paris 2010; 7b: Rue des Archives/PVDE; 8t: Getty Images/Time & Life Pictures/Co Rentmeester; 8b: Collection Christophel/'Star Wars III', Georges Lucas with Ewan McGregor, Silas Carson, 2005, 20th Century Fox; 8/9t: Cosmos/SPL/Roger Harris; 9cl: Cosmos/SPLL/Science Museum; 9cl: © Diners Club; 9b: © McDonald's Corporation; 10t: BSIP/James L Amos; 10bl: Cosmos/SSPL/Science Museum; 10br: Corbis/Jose L Pelaez/JLP; 10/11t: Corbis/Bettmann, 11bl: © Groupe Seb; 11br: Cosmos/SPL/A Barrington Brown; 12tl: Cosmos/SSPL/Science Museum; 12tr: Corbis/Bettmann; 12b: Biosphoto/Peter Arnold/Gordon Wiltsie, 13t: Reuters/Luc Gnago; 13cl: Getty Images/FoodPix/Brian Hagiwara; 13cr: Getty Images/Time & Life Pictures/Francis Miller; 14t: © France Télécom/Michel Le Gal, 14tr: Collection Christophel/'Die Another Day', Lee Tamahori with Pierce Brosnan, 2002, UDF; 14/15b: Leemage/Costa; 15t: The Cousteau Society; 15bl: ® LEGO, the LEGO logo and configuration of LEGO bricks © 2010 The LEGO Group; 15bd: © Courtesy NIST – National Institute of Standards and Technology; 16tl: Cosmos/SSPL/Science Museum; 16bl, 16/17t: Corbis/Bettmann; 16/17cr: Corbis/Bob Sacha; 17l: Corbis/Lester Lefkowitz; 17tr: Leemage/Imagestate; 18/19: Corbis/Bettmann; 20b: AKG-Images/SPL; 20/21t: AKG-Images/SPL; 22t: Corbis/Bettmann; 22b: Collection Christophel/'2001: A Space Odyssey' by Stanley Kubrick, 1968, MGM; 23t: AKG-Images/SPL; 23bl: Cosmos/SSPL/Science Museum; 23br: AKG-Images/SPL ; 24t: © IBM; 24b: Corbis/Saba/Najlah Feanny; 25t: © Cern/Claudia Marcelloni, Geneva; 25b: © With kind permission of Apple; 26tr: Getty Images/Stone/Siegfried Layda; 26cl: Cosmos/SPL/Rutherford Appleton Library/Sheila Terry; 26cr: Cosmos/SSPL/Science Museum; 26bl: Corbis/Lester Lefkowitz; 26br: Bookmaker; 26/27c: AKG-Images/SPL; 27tl & tc: Cosmos/SSPL/Science Museum; 27tr: RÉA/Michel Gaillard; 27bc: BSIP/SPL/South Illinois University; 27b: © With kind permission of Apple; 28t: Leemage/AGF; 28b: Leemage/Costa Archives Milan, 'Roman Holiday' by William Wyler, with Audrey Hepburn, 1953, Paramount Pictures; 29t: Eyedea/Gamma/R.A.; 29b: Collection Christophel/Dogora, 'Ouvrons les yeux', by Patrice Leconte, 2004, Warner Bros; 30t: Corbis/Bettmann; 30cr & b: © Tupperware; 31t: Getty Images/Hulton Archive/Apic; 31c: Cosmos/SSPL/Science Museum; 31b: RÉA/Laif/Joerg Glaescher; 32t: Cosmos/SPL/James King-Holmes; 32b: Getty Images/Time & Life Pictures/J.R. Eyerman; 32/33t: AKG-Images/Erich Lessing/British Museum, London; 33b: Cosmos/SPL/James King-Holmes; 34cl: BSIP/Photo Researchers; 34t: Cosmos/SPL/Lawrence Berkeley Laboratory; 34/35b: Cosmos/SPL/David Parker; 35t: BSIP/Science Source; 36t: Cosmos/SPL/Brookhaven National Laboratory; 37t: © Cern/Maximilien Brice, Geneva; 37b: © Cern Geneva; 38t: Cosmos/SSPL/Science Museum; 38cl: Cosmos/SPL/Emilio Segre Visual Archives/American Institute of Physics; 39: Corbis/Bettmann; 40t: AKG-Images/SPL; 40c: Cosmos/SPL/Tom Kinsbergen; 40b: Cosmos/SPL/Tony Craddock; 41t: The Bridgeman Art Library/DaTo Images, Philips publicity poster, 1952, 'Pierre Masseau dit Fix-Masseau' (1905-1994), private collection © Adagp, Paris 2010; 41b: Corbis/Hulton-Deutsch Collection; 42t: © 2010 Musée du Quai-Branly, Paris, Photo Patrick Gries/Bruno Descoings/Scala, Florence; 42b: Musée de l'Homme – Photothèque © 2010 Musée du Quai Branly, Photo Claude Lévi Strauss/Scala, Florence; 42/43t: Rue des Archives/PVDE; 43b: 'Tristes Tropiques' by Claude Lévi-Strauss, from the author's collection, Univers Poche, 1955; 44cl: Pascal Goetgheluck; 44/45t: Collection Christophel/'Star Wars III' by Georges Lucas with Ewan McGregor, Silas Carson, 2005, 20th Century Fox; 45b: Eyedea/Age/Bilderbox; 46t: Cosmos/SSPL/Science Museum; 46b: Getty Images/Martin Barraud; 47t. Getty Images/Time & Life Pictures/Co Rentmeester; 47b: Corbis/Bettmann; 48b: Cosmos/SPL/Roger Harris; 49cr, 50tl: AKG-Images/SPL; 50tr: Cosmos/SPL/'Physics Today' Collection, American Institute of Physics; 50/51c: Cosmos/SPL/NASA; 51t: Cosmos/SPL/Detlev Van Ravenswaay; 53tl: Ciel et Espace; 53br: Corbis/EPA/NASA; 54bl: Cosmos/SPL/NASA; 55tl: Getty Images/Time & Life Pictures/Carl Iwasaki; 55tr: 'One Two Three ... Infinity' by George Gamow/DR; 55cr: © Le Pommier, 2002; 56tr: Cosmos/SSPL/Science Museum; 56c: Corbis/Bettmann; 56b: Cosmos/SPL/Dr Ken MacDonald; 57t: Ifremer/Éric Lacoupelle; 57c: Corbis/Ralph White; 58tl: Leemage/Selva; 58c: Corbis/Bettmann; 58/59c: Eyedea/Gamma/Piccard Collection; 59cr: Eyedea/Gamma/Francis Demange; 60tl: Getty Images/Time & Life Pictures/Werner Wolff, 60b: © McDonald's Corporation; 61t: RÉA/Pascal Sittler; 61b: © Scott Blake, www.barcodeart.com; 62c: © Diners Club; 62b: Corbis/Blend-Images/Rolf Bruderer; 63t: BSIP/James L. Amos; 63b: Getty Images/Jake Fitzjones; 64bl: Getty Images/Time & Life Pictures/Joel Yale; 64c: Cosmos/SSPL/Science Museum; 65t: Cosmos/SPL; 65tr: Cosmos/SPL/Volker Steger; 66tr: Corbis/Jose L. Pelaez/JLP; 66cl: Getty Images/Time & Life Pictures/Carl Iwasaki; 67t: Corbis/Ecoscene/Alan Towse; 67b: Look at Sciences/Mona Lisa; 68t: AFP/Tass; 68b: Corbis/Bettmann; 69: AFP; 69b: © Groupe Seb; 70/71t: Cosmos/SPL/A. Barrington Brown; 71t: BSIP/Science Source; 71c: Cosmos/SPL; 71b: Cosmos/SSPL/Science Museum; 72: Look at Sciences/Patrick Landmann; 72/73b: Cosmos/SPL/Pasieka; 73tl: Cosmos/SPL/David Mack; 73tr: Cosmos/SPL/Tim Evans; 74b: Collection Christophel/'Jurassic Park III' by Joe Johnston, 2001, United International Pictures; 75t: Corbis/Frans Lanting; 75b: Getty Images/SSPL/Science Museum; 76tr: Corbis/Bettmann; 76cl: Leemage/MP; 77t: SIPA PRESS/Retro; 77b: Biosphoto/Peter Arnold/Gordon Wiltsie; 78/79t: Corbis/Bettmann; 78c: Mindden Pictures/J.H. Editorial/Hedgehog House/Colin Monteath; 79cr: Corbis/Roger Ressmeyer; 79b: Look at Sciences/Karim Agabi; 80tl: Corbis/Rick Price; 80bl: Corbis/Frans Lanting; 80/81t: Minden Pictures/J.H. Editorial/Michio Hoshino; 81b: Biosphoto/J.L. Klein & M.L. Hubert; 82t: Cosmos/SSPL/NMeM; 82bl: Cosmos/SSPL/Science Museum; 82br: Rue des Archives/Agip; 83t: RÉA/Reporters/Michel Gouverneur; 83c: Collection Christophel/'Be Kind Rewind' by Michael Gondry, with Jack Black and Mos Def, 2008, EuropaCorp Distribution; 84: AFP/Ria Novosti/Lev Nosov; 85t: Corbis/Bettmann; 85b: AFP/Getty Images/Patrick Landmann; 86/87t: Corbis/Karen Kasmauski; 86b: Corbis/Pallava Bagla; 87tl: RÉA/Laif/Andreas Fechner; 87tr: Look at Sciences/Patrick Landmann; 88t: RÉA/Frédéric Maigrot; 88b: RÉA/Jean Pottier; 89: RÉA/Mario Fourmy; 90t: Collection Kharbine Tapabor/'La poêle Tefal n'attache pas', 1960, Raymond Savignac (1907-2002) © Adagp, Paris 2010; 90b: Getty Images/FoodPix/Brian Hagiwara; 91t: AKG-Images; 91b: Reuters/Luc Gnago; 92: Getty Images/Time & Life Pictures/Francis Miller; 93t: Eyedea/Keystone; 93b: Look at Sciences/Eurelios/Philippe Plailly; 94l: RÉA/Patrick Allard; 94r: RÉA/Laif/Andreas Hub; 95t: AFP/Philippe Desmazes; 95cr: AFP/Denis Charlet; 95b: Cosmos/SPL/Cordelia Molloy; 96t: Roger-Viollet/TopPhoto; 96c: Getty Images/Popperfoto; 96b: Eyedea/Keystone; 97: BSIP/Platriez; 98/99t: Cosmos/SSPL/Science Museum; 98b: Leemage/Costa; 99t & c: Corbis/Roger Ressmeyer; 100t: AFP/Mod © Crown Copyright; 100b: AFP/François Mori; 101t: Getty Images/Getty Images News; 101b: Corbis/Bettmann; 102t: Getty Images/Time & Life Pictures/Michael Rougier; 102b: © François Maréchal for France Télécom; 103t: © France Télécom/Michel Le Gal; 103b: RÉA/Pierre Bessard; 104l: Sipa Press/Bruno Bebert; 104/105: SeaPics.com/J.H. Editorial/Reinhard Dirscherl; 105tl & bl: The Cousteau Society; 106c: Sipa Press/'Le Monde du Silence' by Jacques-Yves Cousteau and Louis Malle, 1956; 106b, 107tl: The Cousteau Society; 107tr: Sipa Press/Zeppelin; 108tl: Corbis/Bettmann; 108c: Sipa Press/Neco/Bruno Delessard; 108/109t: Getty Images/Panoramic Images; 109br: The Cousteau Society; 110t: Getty Images/SSPL/Science Museum; 110b: Corbis/Hulton-Deutsch Collection; 111t: AFP/Philippe Hugen; 111b: Collection Christophel/'Die Another Day' by Lee Tamahori with Pierce Brosnan, 2002, UDF; 112l: Corbis/Kulka; 112r: Collection Christophel/'Gentlemen Prefer Blondes' by Howard Hawks with Marylin Monroe, 1953, 20th Century Fox Film Corporation; 113t: Corbis/Bettmann; 113b: © Courtesy NIST – National Institute of Standards and Technology; 114t: Cosmos/SPL/Eye of Science; 114bl & br: ® LEGO, the LEGO logo and configuration of LEGO bricks © 2010 The LEGO Group; 115t: Getty Images/Hulton Archive/Constance Bannister Corp; 115cl: Corbis/Hulton-Deutsch Collection; 115b: Corbis/TempSport/Jérôme Prévost; 116t: Cosmos/SSPL/Science Museum; 116b: Cosmos/SPL/Léa Paterson; 117t: Corbis/Bettmann; 117b: Oredia/Ajphoto; 118tl: Eyedea/Rapho/Don Carl Steffen; 118tr: Leemage/Fototeca; 118b: Eyedea/Rapho/Janine Niepce; 119: Sipa Press/AP/Khalil Senosi; 120: Roger-Viollet; 121tl: Cosmos/SPL/Pasieka; 121tr: Cosmos/SPL/Tek Image; 121b: RÉA/Ith/Ford; 122t: Corbis/Hulton-Deutsch Collection; 122b: Oredia/Hop américain/Ajphoto; 123tl, tr & c: © 2010 Mattel, Inc. All rights reserved; 123b: Super Glue Corp®; 124t: Corbis/Lester Lefkowitz; 124b: Getty Images/Time & Life Pictures/Fritz Goro; 125t: Cosmos/ SPL/David Parker Seagate Microelectronics Ltd; 125b: BSIP/Photo Researchers; 126t: Corbis/Bettmann; 126b: Corbis/Richard Hamilton Smith; 127t: Getty Images/Popperfoto; 127b: Eyedea/Keystone; 128t: Rue des Archives/Agip; 128b: Corbis/Bettmann; 129t: Getty Images/Popperfoto; 129cr: Corbis/Bettmann; 130t: Roger-Viollet; 130br: Getty Images/ABC via Getty Images/ABC Photos Archives, 'The Untouchables' with Robert Stack, 1962; 130bl: Corbis/Mascarucci; 131: Rue des Archives/Agip; 132t: Rue des Archives/Imago; 132c: Getty Images/Hulton Archive/CBS Photo Archive; 132b: Eyedea/Age/San Rostro; 133t: Hemis.com/Paule Seux; 133b: AFP/You Tube; 134/135t: Corbis/Bettmann; 134b: Cosmos/SPL/Corning Inc, Emilio Segre Visual Archives, American Institute of Physics; 135t: Cosmos/SSPL/Science Museum; 135b: © NASA/JPL/Space Science Institute; 136/137t: Corbis/Bob Sacha; 136b: Corbis/NASA/Roger Ressmeyer; 137c: Collection Christophel/'Superman III' by Richard Lester with Christopher Reeve, 1983, Warner Bros; 137b: Getty Images/Redferns; 138t: Cosmos/SPL/Simon Fraser; 138b: DoubleVue.fr/Philippe Psaïla; 139t: Cosmos/SPL/Klaus Guldbrandsen; 139b: RÉA/The New York Times; 140: Photononstop/Focus Database/Tips Berlaymont, Années 1960, Architecte Lucine de Vestel DR; 141t: AFP; 141b: AKG-Images/Paul Almasy/Pavillon Philips, 1958, Le Corbusier © FLC/Adagp Paris 2010; 142t: Leemage/Imagestate; 142b: AFP/Ria Novosti ; 143tl: © Citroën Communication; 143tr: Eyedea/Keystone/Architecte Guillaume Gillet; 143b: Hemis.com/Emilio Suetone; 144/145: AKG-Images/SPL; 146tl: Cosmos/SSPL/Science Museum; 146tr: Getty Images/Hulton Archive/Apic; 146b: © IBM; 147tr: Cosmos/SSPL/Science Museum; 147b: Leemage/Costa Archives Milan, 'Roman Holiday' by William Wyler with Audrey Hepburn, 1953, Paramount Pictures; 147br: © Tupperware; 148bl: Cosmos/SSPL/Science Museum; 148br: 'One Two Three ... Infinity' by George Gamow/DR; 148/149b: Cosmos/SPL/Pasieka; 149tl & cl: The Cousteau Society; 149tr: © Scott Blake, www.barcodeart.com; 150tl: Corbis/Kulka; 150bl: Getty Images/Hulton Archive/Constance Bannister Corp; 150br: Corbis/Hulton-Deutsch Collection; 151tl: © 2010 Mattel, Inc. All rights reserved; 151tc: Super Glue Corp®; 151tr: Getty Images/Popperfoto; 151b: Corbis/Hulton-Deutsch Collection; 152cl: Eyedea/Keystone; 152b: AKG-Images/Paul Almasy/Pavillon Philips, 1958, Le Corbusier © FLC/Adagp Paris 2010; 152tr: Leemage/Selva; 153t: Corbis/Bettmann; 153bl: Cosmos/SPL/Simon Fraser; 153r: Roger-Viollet/TopPhoto.

Illustrations on page 36/37 (Large Hadron Collider), 44 (bottom, producing a hologram), 48/49 & 148t (after the Big Bang), 52 (from the Big Bang to the present) – by Grégoire Cirade.

THE ADVENTURE OF DISCOVERIES AND INVENTIONS
The Age of Big Science – 1945 to 1960
Published in 2011 in the United Kingdom by Vivat Direct Limited
(t/a Reader's Digest), 157 Edgware Road, London W2 2HR

Adapted from *Les Inventions Pour Tous*, part of a series entitled L'ÉPOPÉE DES DÉCOUVERTES ET DES INVENTIONS, created in France by BOOKMAKER and first published by Sélection du Reader's Digest, Paris, in 2010.

Translated from French by Peter Lewis

PROJECT TEAM
Series editor Christine Noble
Art editor Julie Bennett
Designer Martin Bennett
Consultant Ruth Binney
Proofreader Ron Pankhurst
Indexer Marie Lorimer

Colour origination FMG
Printed and bound in China

VIVAT DIRECT
Editorial director Julian Browne
Art director Anne-Marie Bulat
Managing editor Nina Hathway
Picture resource manager Sarah Stewart-Richardson
Technical account manager Dean Russell
Product production manager Claudette Bramble
Production controller Sandra Fuller

We are committed both to the quality of our products and the service we provide to our customers. We value your comments, so please feel free to contact us on 0871 3511000 or via our website at **www.readersdigest.co.uk**

If you have any comments or suggestions about the content of our books, you can email us at **gbeditorial@readersdigest.co.uk**

CONCEPT CODE: FR0104/IC/S
BOOK CODE: 642-011 UP0000-1
ISBN: 978-0-276-44523-1